梁思成图说西方建筑

Illustrated Notes on Western Architecture by Liang Sicheng

梁思成 著 林洙 编

外语教学与研究出版社
FOREIGN LANGUAGE TEACHING AND RESEARCH PRESS
北京 BEIJING

梁思成说：

"要做笨人，下笨功夫."

　　　　　林洙 2013.11.26.

EDITOR'S NOTE
出版说明

《梁思成图说西方建筑》一书是梁思成先生1924年赴美国宾夕法尼亚大学建筑学院读书期间，修读西方建筑史课程时所做的读书笔记和课堂作业的汇编。笔记原稿已完全散乱。林洙先生根据回忆，将这些凌乱的笔记重新梳理，遗憾的是笔记原来的排序已经无从可考。为了读者赏读方便，本书以梁思成先生修读这门课时教师的指定参考书《弗莱彻建筑史》（*Sir Banister Fletcher: A History of Architecture on the Comparative Method for the Student, Craftisman and Amateur*）为主要参考依据，以教师课堂布置的及课外临摹的手绘图为正文主线，将正文分为两部分：第一部分从建筑史流派的角度将书稿分为意大利罗马式建筑、哥特式建筑及文艺复兴时期的建筑三个章节；第二部分将无法按照学派划分的手绘图和文字摘要按地域划分。为了向读者更加真实、全面地展现梁思成先生求学时所下的苦功，本书亦将先生根据教授讲授内容随堂做的文字笔记收入书中，以附录呈现，使读者更加深入的了解先生的刻苦学习精神。

为展现原作的时代特点并尊重历史原貌，笔记按照原稿大小真实呈现，对笔记原稿中出现的行文讹误不作修改。为方便读者深入理解文本，将笔记中的英文内容翻译为中文，并依据相关专业资料，对作者提及的主要建筑师及未予文字详述的建筑物作了简要注释。

FOREWORD
前言

1953 年，我从重工业部调到清华大学建筑系工作。当时的系馆设在清华学堂大楼内，一进系馆，正面是一个很宽的老式木楼梯，楼梯的两侧是宽宽的走廊。右面走廊靠墙的一面依次排列着西方古典建筑的五种柱式，走廊靠窗的一面依次排列着中国各时期斗拱的模型。西方古典柱式是白色的，与棕色硬木的斗拱呈现出鲜明的对比。走廊尽头是系图书馆。左面的走廊靠窗一侧放着一条长案，上面排列着"样式雷"的彩色模型，尽头是系资料室。走上楼梯，对面是一个大陈列室，里面布置着各式各样的明清家具。这个系馆给了我强烈的"建筑意"与浓厚的历史文化感。

1958 年，我调到系资料室工作。一天，我在一大堆废书籍中发现了几个笔记本，它们的封面已弄得很脏*，看不出模样，打开一看竟全是英文打字的文稿，而且每隔几页就有一小幅精美的小钢笔画。我惊呆了，再仔细一看，在画的一旁往往有一个钢印，上面写着宾夕法尼亚大学建筑学院的字样。宾夕法尼亚大学不正是梁先生的母校吗？于是我拿着这几个笔记本找到梁公，请他看看是否是他的东西。梁公接过来一看，脸上微微露出一副难以捉摸的表情，轻声地说："是的，这是我学西方建筑史时做的笔记。"我说："您是否要收回呢？"他笑了笑说："既然现在在你处，那就交给你吧！"我听罢如获至宝，从此我便常常翻阅这些笔记。那些精美的小画是怎样画出来的呢？一门课就画了一百多幅画，使我肃然起敬。

后来我又听到一个故事。原来按宾大的教学计划，建筑史是安排在二年级才学的课程，梁思成因受父亲的影响，对历史特别感兴趣，就自己跑去旁听，越听越感兴趣，于是他找到建筑史的教授请求允许他提前一年选修此课。授课的教授对这个热爱建筑史的学生非常有好感，同意了他的请求。

我细细地翻看着这些笔记。他是一个多么用功的学生啊！除了详细地记录老师讲授的内容外，他还阅读了大量专业著作并做了笔记。《弗莱彻建筑史》是教师指定的参考书，除了教师规定的作业外，他还从其他书中临摹了不少作品。难怪当年先生讲授西方建筑史时，能准确地在黑板上勾画出很多著名建筑

* 1952年我国开始全面学习苏联，建筑学也采用了苏联的教材，因为冷战的影响，中美关系在很长一段时间内处于敌对状态，梁先生在美国学习时的笔记本也被扔进了垃圾堆。

物的平立剖面图。正因为他对西方建筑史下过苦功夫，所以他对东西方建筑的差别极其敏感，而这也是他日后研究中国建筑史的重要基础之一。

梁思成在美国学习时，深深感到西方国家对本国的建筑史非常重视，而且运用先进的科学技术进行研究整理，并写出本国的建筑史。回顾我国，尽管历史悠久，却没有一部建筑史，因此他决心要研究中国的建筑发展史。

这里需要向读者说明的是，本书并不是梁思成的学术著作，仅仅是他初到美国在一年级时学习建筑史课的笔记。当时他刚刚结束了清华学校留美预备班的学业，所以不管是英语还是有关欧洲历史文化方面的知识都还只是个初学者的水平，对欧洲中世纪的建筑师、雕塑家、画家的了解更是微乎其微。这个笔记本只是教师讲课的记录，及在课外参考书中查阅到的有关著名建筑师的生平及他们的作品……

在中世纪建造一座巨型的建筑往往要经历一二百年甚至更长的时间，跨越两三代王朝，凭借众多名建筑师的努力才能完成。这些内容在参考书中均有丰富的记述，他也就原样抄录下来，作为刚入门的学生，他还没有能力去考证这些材料的准确性，更不可能对收集到的材料作出系统的整理。再者，这些90年前的笔记，在建筑词汇及语法上与现代语言也会有些差异，尽管外研社的朋友们付出很大的努力，但读起来仍感不畅与晦涩，这是要请读者谅解的。

但这些笔记反映了我国老一辈学者是怎样踏进了现代建筑学的领地，反映了老一辈学者坚毅刻苦的精神，也是中西方文化交流的一个实物见证，正如外研社吴浩先生说的，"它们不仅仅是笔记，更是一份珍贵的历史文献"。

"文革"期间，我家数度被抄，这几个笔记本再次遭受厄运，而今我又一次把它们整理出来，但已缺损，难以恢复原貌。作为90年前一个普通学生的笔记本，它们得以保存至今的确不易。我谨将它们作为一份珍贵的文献，奉献给读者。

最后，我非常感谢外研社吴浩先生和张昊媛女士。他们对本书的出版付出了极大的努力，并在质量上精益求精。为了弄清一座建筑的正确译名，张女士参阅了很多资料并多次求教于王瑞珠院士。我还要特别感谢设计师潘振宇、高瓦夫妇，他们一边照顾着刚出生的小女儿，一边考虑怎样把这本书做得更完美。这一切都使我感动，并深深致谢。

林洙
2015.9.25

1924—1927年，梁思成就读于美国宾夕法尼亚大学建筑学院。我国近代建筑大师中的童寯、陈植、杨廷宝也先后在此就读。据梁先生的大学同窗陈植先生回忆："在宾大，思成兄就学期间全神以赴、好学不倦给我以深刻的印象。我们常在交图前夕彻宵绘图或渲染，他是精益求精，我则在弥补因经常欣赏歌剧和交响乐而失去的时间。在当时'现代古典'之风盛行的影响下，思成兄在建筑设计方面鲜落窠臼，成绩斐然，几次评为一级。他的设计构图简洁，朴实无华，亦曾尝试将建筑与雕塑相结合，以巨型浮雕使大幅墙面增添风韵。他的渲染，水墨清澈，偶用水彩，则色泽雅淡，明净脱俗。除建筑设计外，思成兄对建筑史及古典装饰饶有兴趣，课余常在图书馆翻资料，做笔记，临插图，在掩卷之余，发思古之幽情……"右图为梁思成（右）与陈植在美国宾夕法尼亚大学的合影。

Membership Certification of Tau Sigma Delta of Honorary Fraternity
in Architecture and Allied Arts at the University of Pennsylvania
May 2nd, 1927

Tau Sigma Delta (ΤΣΔ) 建筑与相关艺术荣誉联谊会
宾夕法尼亚大学Epsilon分会入会证书
1927年5月2日

THE ARCHITECTURAL SOCIETY

OF THE UNIVERSITY OF PENNSYLVANIA

HEREBY CERTIFIES THAT

LIANG SHIH-CHENG

HAS BEEN ELECTED TO MEMBERSHIP IN THE SOCIETY AND IS THEREFORE ENTITLED TO ENJOY ITS PRIVILEGES AND EXPECTED TO SHARE ITS RESPONSIBILITIES.

OCTOBER 1927.

John N. Richards   *Henry Louis Sandlass*
SECRETARY   **PRESIDENT**

Certification of the Architectural Society of the University of Pennsylvania
October 1927

宾夕法尼亚大学建筑学会入会证明
1927年10月

Praeses et Curatores Vniuersitatis Princetoniensis
Omnibus has litteras lecturis
Salutem in Domino.

Quandoquidem aequum sit et rationi prorsus consentaneum eos qui labore et studio bonas artes didicerunt praemia suis meritis digna referre, eo ut et ipsis bene sit et aliorum prouocetur industria, quando etiam huc potissimum spectant amplissima illa iura nostrae Vniuersitati publico diplomate collata.

Quumque **Liang Ssu-ch'eng** air sit non tantum moribus inculpatus, litteris humanioribus penitus instructus, sed etiam sibi tantam in **architectura Serica colenda** cognitionem acquisiuerit ut summos publicos honores probe mereatur:

Idcirco notum sit omnibus quod nos Senatusconsulto Academico supradictum **Liang Ssu-ch'eng** titulo graduque **Litterarum Doctoris** honoris causa adornandum et dehinc pro Doctore habendum uoluimus.

Cuius rei haec membrana sigillo nostrae Vniuersitatis rata et nominibus Praesidis et Scribae manuta testimonio sit.

Datum Aulae Nassouicae
die III Aprilis
Anno Domini MDCCCXLVII

Praeses

Scriba

Certification of Honorary Doctor of Letters at Princeton University in 1947

普林斯顿大学荣誉文学博士证书，1947年

CONTENTS
目录

PART I
第一部分

ARCHITECTURE NOTES: A HISTORICAL PERSPECTIVE
建筑笔记：建筑史流派的视角

ITALIAN ROMANESQUE 意大利的罗马式建筑・2

- Basilica of Saint Paul outside the Walls, Rome
  城外圣保罗教堂，罗马・2

- Basilica di San Michele Maggiore, Pavia
  圣米凯莱教堂，帕维亚・3

- The Clock Tower, Venice
  圣马可时钟塔，威尼斯・4

ITALIAN GOTHIC 意大利的哥特式建筑・5

- Palazzo Vecchio, Florence
  旧宫，佛罗伦萨・5

- Certosa di Pavia
  帕维亚修道院・6

- Basilica of Saint Anthony, Padua
  圣安东尼教堂，帕多瓦・7

- Court of the Doge's Palace, Venice
  威尼斯总督府・8

ITALIAN RENAISSANCE ARCHITECTURE 意大利文艺复兴时期的建筑 · 10

The Florentine School 佛罗伦萨学派 · 10

Brunelleschi 布鲁内莱斯基 · 10

- The Dome of the Cathedral of Florence 佛罗伦萨大教堂穹顶 · 14
- The Pazzi Chapel, Florence 帕齐礼拜堂，佛罗伦萨 · 15
- Basilica of San Lorenzo, Florence 圣洛伦佐教堂，佛罗伦萨 · 17
- Santo Spirito, Florence 圣灵教堂，佛罗伦萨 · 18
- The Palazzo Pitti, Florence 皮蒂府邸，佛罗伦萨 · 19
- Ospedale degli Innocenti, Florence 孤儿院，佛罗伦萨 · 21
- Palazzo Pazzi-Quaratesi, Florence 巴齐－夸拉泰西府邸，佛罗伦萨 · 23

Alberti 阿尔贝蒂 · 25

- San Francesco, Rimini 圣弗朗切斯科教堂，里米尼 · 25
- Santa Maria Novella, Florence 新圣马利亚教堂，佛罗伦萨 · 26
- Basilica of Sant'Andrea, Mantua 圣安德烈教堂，曼图亚 · 27
- Window in Palazzo Strozzi, Florence 斯特罗兹宫窗户，佛罗伦萨 · 29

The Roman School 罗马学派 · 30

Bramante 布拉曼特 · 30

- Santa Maria delle Grazie, Milan 圣马利亚感恩教堂，米兰 · 33
- Palazzo della Cancelleria, Rome 坎榭列利亚宫，罗马 · 40
- Vatican 梵蒂冈宫 · 54
- IL Tempietto, Rome 坦比哀多礼拜堂，罗马 · 56
- Santa Maria della Pace, Rome 和平圣马利亚教堂，罗马 · 57
- St. Peter's Basilica, Rome 圣彼得大教堂，罗马 · 59
- Santa Maria presso San Satiro, Milan 圣沙弟乐圣母堂，米兰 · 75

Michelangelo 米开朗琪罗 · 79

- The Capital, Rome 卡皮托利尼博物馆，罗马 · 79

The Venetian School 威尼斯学派 · 80

Palazzo Vendramin, Venice 文德拉明宫，威尼斯 · 80

Scuola Grande di San Marco, Venice 圣马可学校，威尼斯 · 82

Santa Maria dei Miracoli, Venice 奇迹圣母堂，威尼斯 · 85

Palazzo Corner Spinelli, Venice 科纳·斯皮内利府邸，威尼斯 · 90

San Zaccaria, Venice 圣匝加利亚教堂，威尼斯 · 91

Palazzo del Consiglio, Verona 议会大厦，维罗纳 · 93

Palazzo Bevilacqua, Bologna 饮泉宫，博洛尼亚 · 94

Palazzo Fava Ghisilieri, Bologna 法瓦大殿，博洛尼亚 · 98

The Lombard School 伦巴第学派 · 100

Filarate 菲拉雷特 · 102

- The Ospedale Maggiore, Milan 米兰市立医院 · 105

PART II
第二部分

ARCHITECTURE NOTES: A GEOGRAPHICAL PERSPECTIVE
建筑笔记：地域的视角

ROME 罗马 · 108
- Palazzo Venezia & Church of S. Mark 威尼斯宫和圣马可教堂 · 108
- Santa Maria dell'Anima 圣马利亚灵魂之母教堂 · 113
- Santa Maria del Popolo 人民圣母教堂 · 118
- San Pietro in Montorio 蒙托利尔的圣彼得教堂 · 120
- Trajan's Column 图拉真柱 · 122

MILAN 米兰 · 123
- The Door of the Medici Bank 美第奇银行的门 · 123
- Basilica of San Lorenzo 圣洛伦佐教堂 · 124
- Abbiategrasso 阿比亚泰格拉索教堂 · 125
- Chapel of San Pietro Martire 殉道者圣彼得小教堂 · 126

FLORENCE 佛罗伦萨 · 127
- Villa Careggi 卡勒吉别墅 · 127
- Santissima Annunziata 圣母领报大殿 · 128
- Palazzo Medici Riccardi 美第奇－里卡迪宫 · 129
- Basilica of Santa Croce 圣十字教堂回廊 · 132
- The Marsupini Tomb in S. Croce 圣十字教堂马苏匹尼基 · 133
- Chapel of Crucifix（San Miniato）圣体小堂（圣米尼亚托教堂）· 134

VENICE 威尼斯 · 135
- Procuratie Vecchie 旧行政长官官邸大楼 · 135
- Chiesa di San Giovanni Evangelista 圣约翰福音教堂 · 136
- Scuola Grande di San Rocco 圣洛可大会堂 · 137
- Palazzo Dairo 达里欧宫 · 140

BOLOGNA 博洛尼亚 · 141
- Church of San Spirito 圣灵教堂 · 141
- Palazzo Carracci 卡拉奇宫 · 142
- Palazzo Pallavicini 帕拉维奇尼宫 · 143
- Palazzo degli Strazzaroli 斯特拉察罗里宫 · 144
- Corpus Domini 圣体教堂 · 145

VERONA 维罗纳 · 146
- English Gothic Ornaments 英国哥特式装饰 · 146

BRESCIA 布雷西亚 · 147
- Palazzo Communale 市政府 · 147
- Santa Marie dei Miracoli 奇迹圣母堂 · 148
- Monte di Pietà 典当行 · 149

PADUA 帕多瓦 · 150
- Loggia del Consiglio 市政会凉廊 · 150

FERRARA 费拉拉 · 151
- Palazzo Diamanti 钻石宫 · 151
- Palazzo Schifanioa 席法诺亚宫 · 152
- Palazzo Roverella 罗维戈宫 · 153
- Palazzo Costabili 科斯塔比利宫 · 154

PERUGIA 佩鲁贾 · 155
- Palazzo Podesta 行政长官官邸 · 155
- Porta di San Pietro 圣彼得门 · 156

PIENZA 皮恩札 · 157
- Palazzo del Pretorio 法庭宫 · 157
- IL Duomo di Pienza 皮恩札主教堂 · 158
- Palazzo Picolomini 皮克罗米尼宫 · 159

MANTUA 曼图亚 · 161
- Chiesa di San Sebastiano 圣塞巴斯蒂亚诺 · 161

OTHERS 其他 · 162
- Greek Theatre 希腊剧场 · 162
- Chateau de Blois, Paris 布卢瓦城堡，巴黎 · 163
- Saint-Eustache, Paris 圣厄斯塔什教堂，巴黎 · 164
- Chateau de Chambord, Paris 香波尔城堡，巴黎 · 165
- The Colonnade of the Louvre, Paris 卢浮宫柱廊，巴黎 · 167
- Chateau de Fontainebleau, Paris 枫丹白露宫，巴黎 · 168
- The Pantheon, Paris 先贤祠，巴黎 · 169
- Choragic Monument of Lysicrates, Athens 奖杯亭，雅典 · 170
- Temple at Philae 菲莱神庙 · 171
- San Sophia, Constantinople 圣索菲亚大教堂，君士坦丁堡 · 172
- Palazzo Picolomini, Siena 皮克罗米尼宫，锡耶纳 · 173

```
APPENDIX
```
附录
```
MANUSCRIPT OF LIANG'S IN-CLASS
NOTES ON HISTORY OF ARCHITECTURE
```
梁思成建筑史课堂笔记原稿 · 175

PART I

第一部分

Architecture Notes:
A Historical Perspective

建筑笔记：建筑史流派的视角

BASILICA CHURCH
OF
ST. PAUL 巴西利卡*式圣保罗教堂
ROME. 罗马

此处①为教授给学生作业的判分，分数为5分制，但与中国5分制判分不同的是，1为最高分，5为最低分。

*巴西利卡，长方形会堂（教堂），古罗马的一种公共建筑物，用作市场、法院和会议大厅。平面为长方形，中有两排列柱或柱墩，并在一端或两端有半圆形的凹室，供法官用。小型的巴西利卡用列柱分为中堂和两侧堂，一端有半圆形凹室。公元313年君士坦丁大帝正式承认基督教为合法宗教后，早先基督教教堂就采用了巴西利卡的形式。古罗马的巴西利卡成为后来所有西方教堂建筑发展的基础。

ITALIAN ROMANESQUE 意大利的罗马式建筑
Basilica of Saint Paul outside the Walls, Rome
城外圣保罗教堂*，罗马 •因位于意大利罗马城墙外南边而得名，是罗马的早期基督教长方形大教堂，建于386年，1823年焚毁，1854年重建完成。凯旋门上有早期基督教镶嵌细工杰作，制于5世纪。回廊建于13世纪。是罗马天主教的四座特级宗座圣殿之一。

S. Michele, Pavia. 圣米凯莱教堂，帕维亚

Representing the Italian Romanesque. 意大利罗马式建筑的代表

LOMBARD SCHOOL. 伦巴第学派（参见100—101页）

Basilica di San Michele Maggiore, Pavia
圣米凯莱教堂，帕维亚

The Clock Tower, Venice
圣马可时钟塔,威尼斯

DOORWAY IN PALAZZO VECCHIO 旧宫门廊
FLORENCE 佛罗伦萨

— (BAUKUNST) —
—（建筑）—

ITALIAN GOTHIC
意大利的哥特式建筑
Palazzo Vecchio, Florence
旧宫，佛罗伦萨

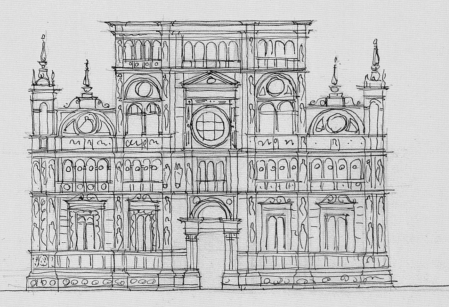

CERTOSA AT PAVIA. 帕维亚修道院

—（BAUM）—
—（鲍姆）—

Certosa di Pavia
帕维亚修道院

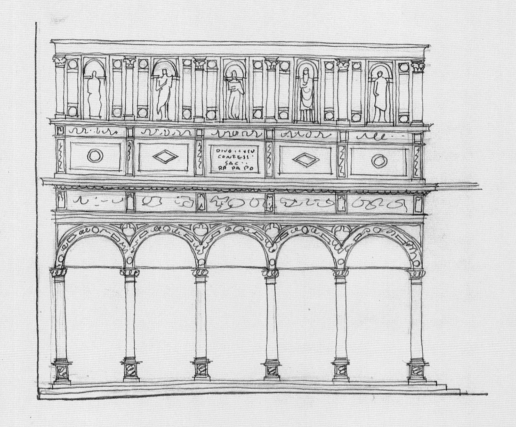

S. ANTONIO,

CAPELLA DEL SANTO 圣安东尼教堂

PADUA. 帕多瓦

—(BAUM - PHOTO) —
—（鲍姆—照片）—

Basilica of Saint Anthony, Padua
圣安东尼教堂，帕多瓦

Court of the Doge's Palace, Venice
威尼斯总督府

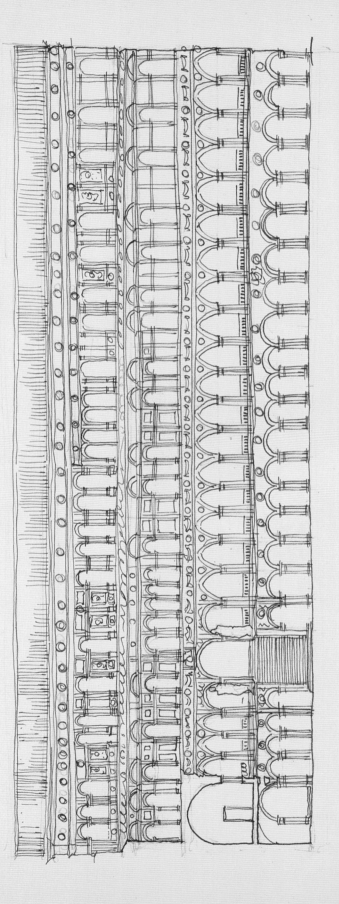

THE MAIN FACADE, COURT OF DOGES PALACE
总督府主立面

ITALIAN RENAISSANCE ARCHITECTURE
意大利文艺复兴时期的建筑

The Florentine School
佛罗伦萨学派

Brunelleschi 布鲁内莱斯基

BRUNELLESCHI

Filippo di ser Brunellescho

Dates and Place

1377 - April 15, 1446   (Florence) - Muntz & Simpson

1375 - 1444 or 1446   - Worthington

Family

Father - Lippo di Turra, notary, minister of war.

Mother - Juliana di Giovanni.

Early Life

Undertook some diplomatic mission after trying law and medicine. Apprenticed to gold smith.   Worked in mechanics, made clocks. 1421 invented ferry over Arno, and mechine for throwing stones, ect., etc.   Tried sculpture.   1401 competition for second baptistry doors of Florence.   Refused commision with Ghiberti. Fabriezy gives 1403 for first trip to Rome, second visit 1405. Manetti places first trip to Rome 1401-1404. — (VASARI)

"At Rome He remained four years; some authotities say wthout a break, others that he returned to Florence for a few months after a year and a half's sojourn.   It is immaterial wiich is true; all agree that the period he spent studying in Rome was considerable   During that time he supported himself partly on the proceeds of a property he had sold before leaving Florence, and partly by working at his craft of sculptor-goldsmith. Vasari givis a vid picture of the life he led there.   He measured and drew the plans and construction of all kinds of buildings, 'temples, round, square, or octagon, basilicas, aquedcuts, baths, arches, the Colosseum, amphitheatres, and every church built of

布鲁内莱斯基

菲利波·布鲁内莱斯基

所在地与对应时间

1377 – 1446 年 4 月 15 日（佛罗伦萨）——芒茨与辛普森

1575 – 1444 或 1446 年——沃辛顿

家庭

父亲——利波·图拉，律师，军务大臣

母亲——朱丽安娜·乔瓦尼

早年生平

在初涉法律和医学之后，曾担任过外交使节。拜金匠为师。工于机械和钟表。*1421 年发明了用于阿诺河的渡轮，另有投石器等发明。尝试过雕塑。1401 年参与竞标佛罗伦萨第二座洗礼堂的大门。拒绝了吉贝尔蒂的委托。1403 年法布里齐第一次去罗马，第二次造访罗马是 1405 年。1401 – 1404 年，马内蒂第一次前往罗马。*——（瓦萨里*）

"他在罗马待了四年；一些权威称，他一直待在罗马；另有观点认为他在逗留一年半之后又返回佛罗伦萨数月。事实究竟如何并不重要，但所有人一致认为他在罗马的学习十分关键。在那段时间，他的资金来源一部分是靠在离开佛罗伦萨前变卖的一处房产，另一部分是靠他做雕塑师和金匠的手艺。"

瓦萨里为他在那里的生活勾勒了一幅生动的图画。他测量并绘制了各种各样建筑、庙宇（圆形、正方形、八角形的）、长方形教堂、高架渠、浴室、拱门、罗马圆形大剧场、半圆形露天剧场和几乎每一座教堂的平面和结构图。

---

* Giorgio Vasari，乔尔乔·瓦萨里 (1511 – 1574)，文艺复兴时期意大利画家和建筑师，以传记《绘画、雕塑、建筑大师传》留名后世。

Brunelleschi

<u>List of Works</u>

Dome of the Cathedral - Florence

Pazzi Chapel - Florence

Sacristry of S. Lorenzo - Flornce

S. Lorenzo - Florence

S. Spirito - Florene

Badia de Fiesole

S. M. Degli Angeli

Ospedale degli Innocenti, Florence

Loggia of th Hospital S. Paolo

Pitti Palace Florence

Palazzo Quaratesi, Florence

Second Cloister of S. Croce, Florence

布鲁内莱斯基

主要作品列表

大教堂穹顶，佛罗伦萨

帕齐礼拜堂，佛罗伦萨

圣洛伦佐教堂老圣器室，佛罗伦萨

圣洛伦佐教堂，佛罗伦萨

圣灵教堂，佛罗伦萨

菲埃索莱大教堂

天使圣马利亚圆堂

育婴院*，佛罗伦萨

圣保罗医院走廊

皮蒂府邸，佛罗伦萨

夸拉泰西*府邸，佛罗伦萨

圣十字教堂第二个柱廊院，佛罗伦萨

* 又名帕齐－夸拉泰西府邸（Palazzo Pazzi-Quaratesi），据资料记载该建筑并非布鲁内莱斯基的作品，而出自另一位意大利建筑设计师朱利亚诺·达·马亚诺基（Giuliano da Maiano）之手

The Dome of the Cathedral of Florence 佛罗伦萨大教堂*穹顶

*又称圣母百花大教堂，一般称为"主教座堂"，是意大利佛罗伦萨的主教堂。

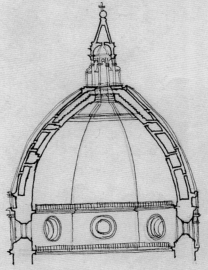
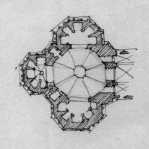

THE DOME 佛罗伦萨大教堂穹顶
of THE CATHEDRAL OF
FLORENCE

Pazzi Chapel, Florence 帕齐礼拜堂，佛罗伦萨

Brunelleschi 布鲁内莱斯基

*图片中有印章的为学生的必交作业。

The Pazzi Chapel, Florence 帕齐礼拜堂，佛罗伦萨

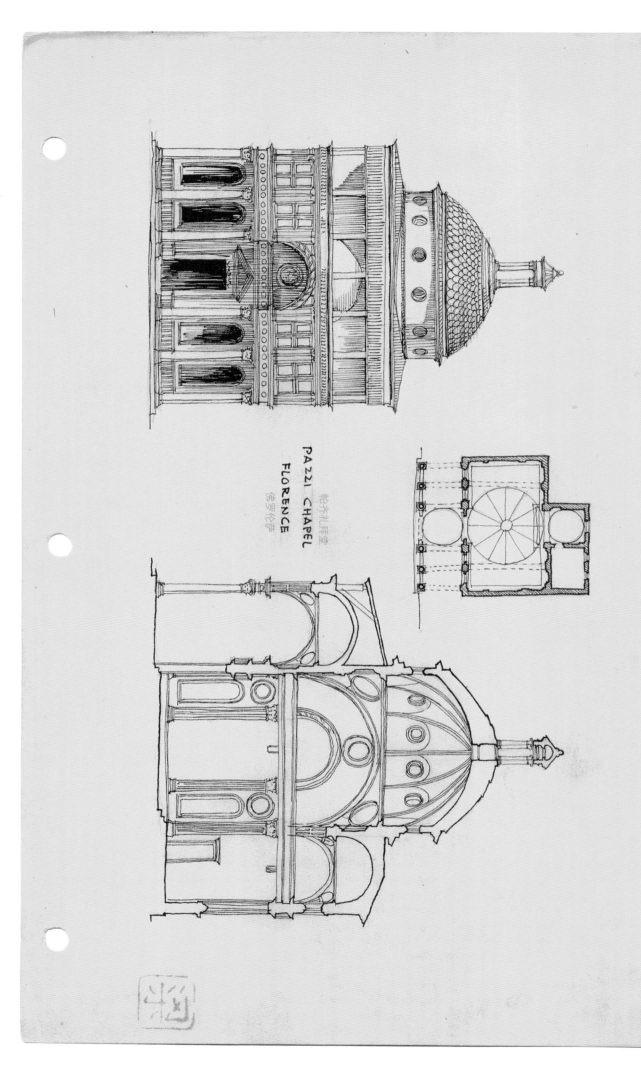

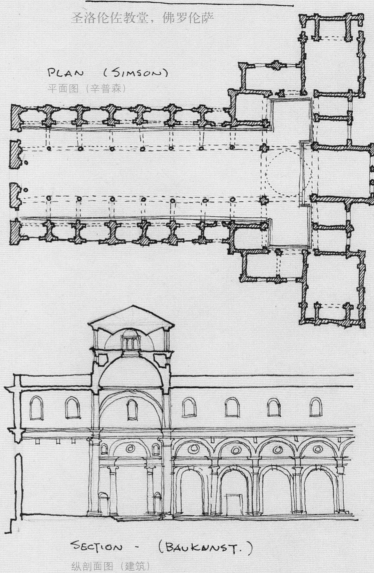

Basilica of San Lorenzo, Florence 圣洛伦佐教堂，佛罗伦萨

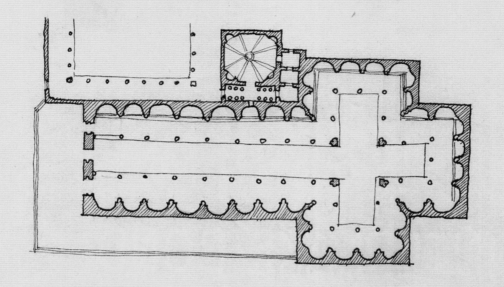

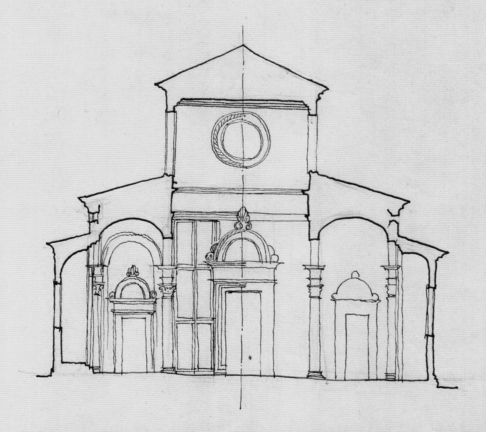

S. SPIRITO, FLORENCE 圣灵教堂，佛罗伦萨
(LASPEYRES)（拉斯拜尔）

Santo Spirito, Florence 圣灵教堂*，佛罗伦萨
*亦称佛罗伦萨圣神大殿。

The Palazzo Pitti, Florence 皮蒂府邸，佛罗伦萨

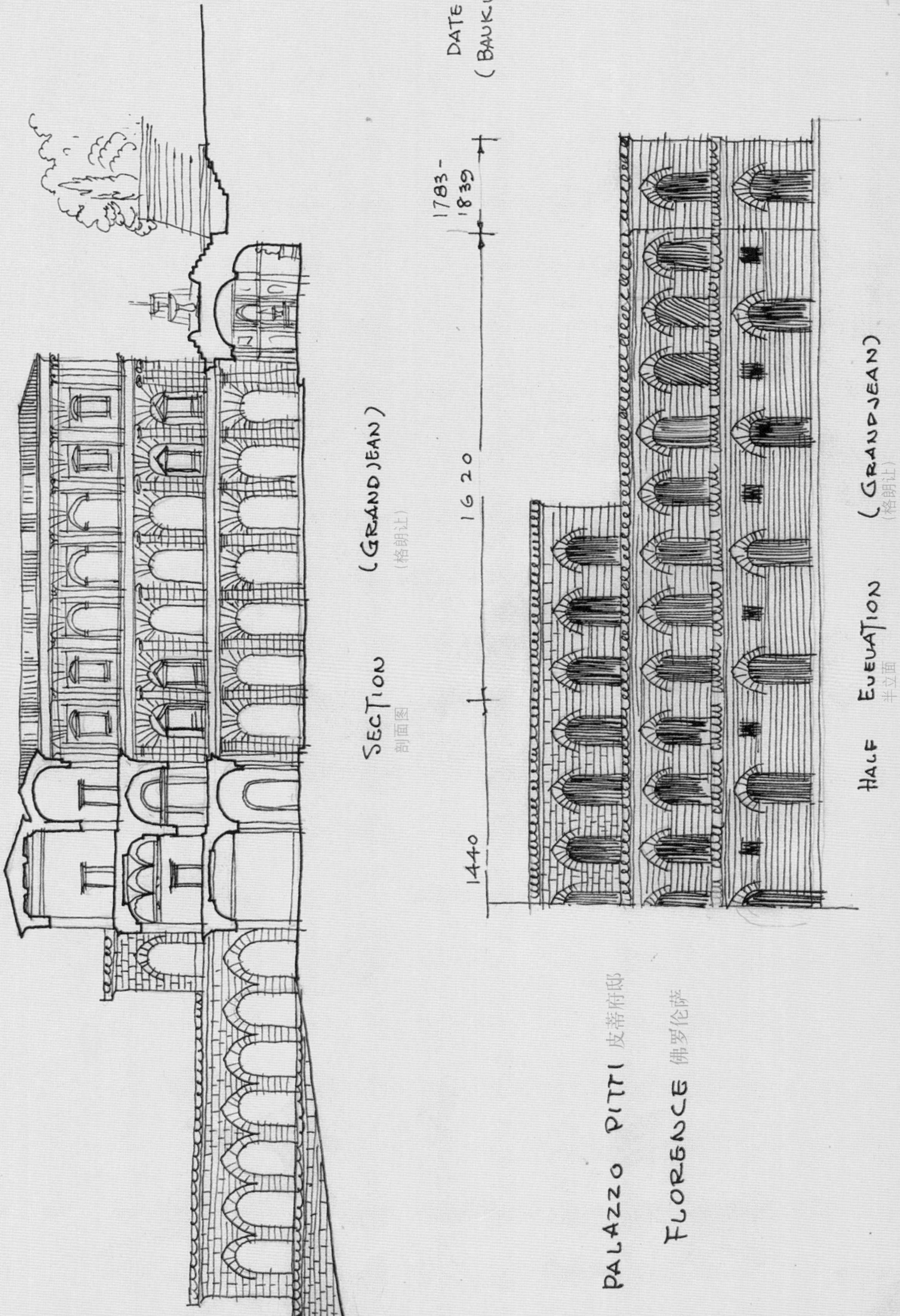

DATES 日期
(BAUKUNST)
(建筑)

1783-1839

1620

1440

SECTION (GRANDJEAN)
剖面图 (格朗让)

HALF ELEVATION (GRANDJEAN)
半立面 (格朗让)

PALAZZO PITTI 皮蒂府邸
FLORENCE 佛罗伦萨
(格朗让)

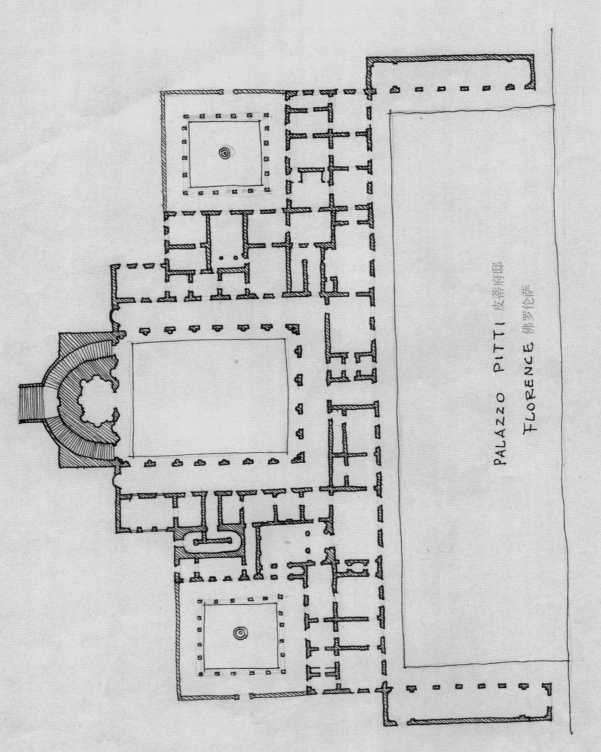

PALAZZO PITTI 皮蒂府邸
FLORENCE 佛罗伦萨

PLAN (GRANDJEAN) 平面图 (格朗让)

Ospedale degli Innocenti, Florence 孤儿院，佛罗伦萨

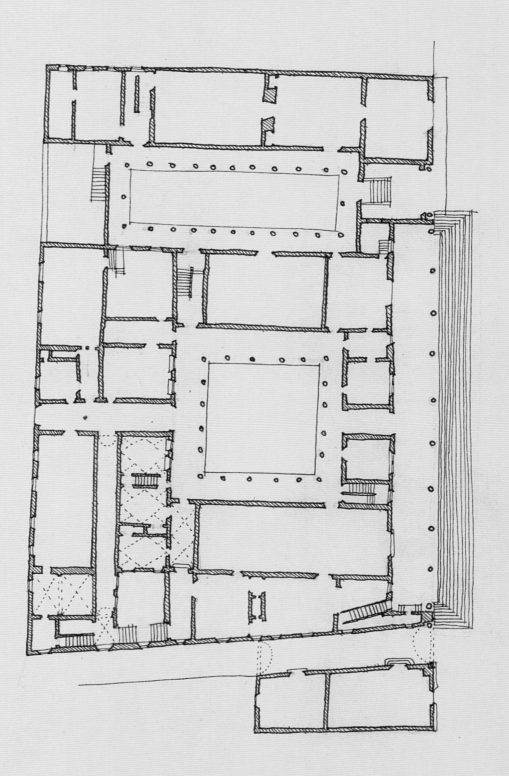

OSPEDALE DEGLI INNOCENTI 孤儿院
FLORENCE 佛罗伦萨

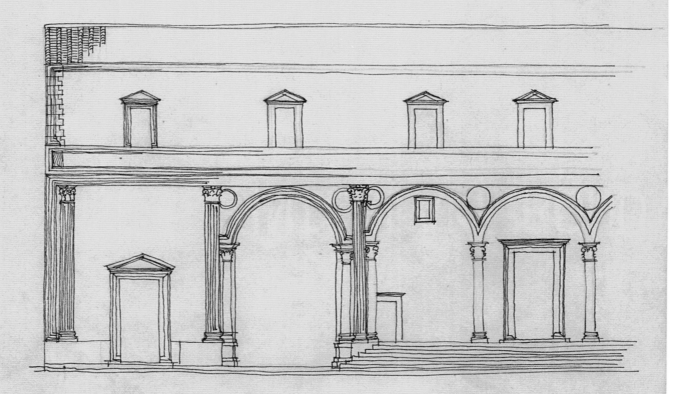

LOGGIA 游廊
OSPEDALE DEGL' INNOCENTI, FLORENCE. 孤儿院,佛罗伦萨

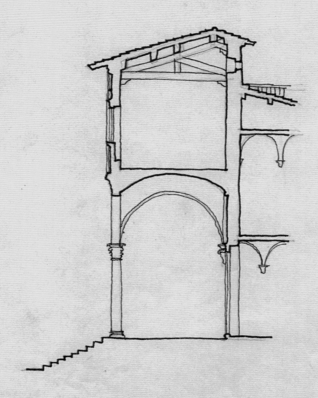

Palazzo Pazzi-Quaratesi, Florence 巴齐-夸拉泰西府邸，佛罗伦萨

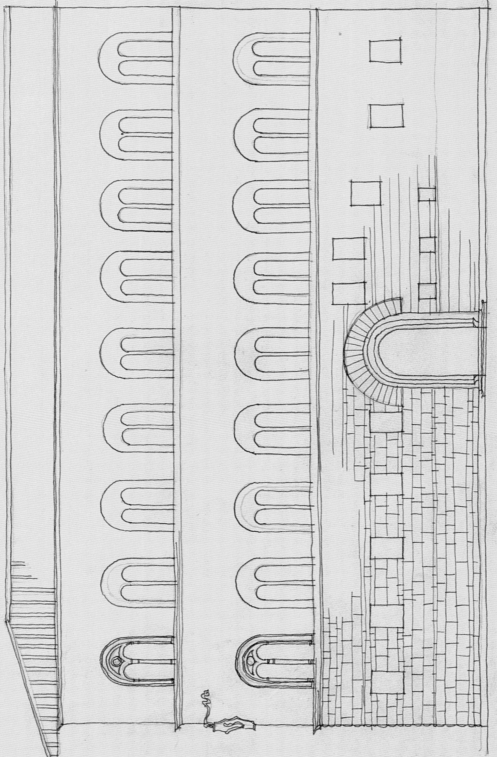

PALAZZO QUARATESI
巴齐-夸拉泰西府邸

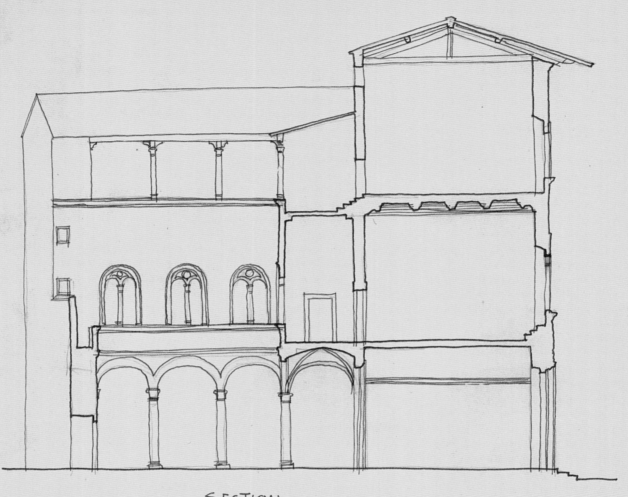

SECTION
剖面图

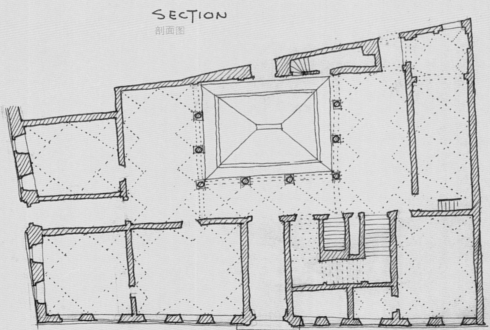

PLAN 平面图

PALAZZO QUARATESI 巴齐—夸拉泰西府邸

FLORENCE 佛罗伦萨 (TOSCANA)(托斯卡纳)

Alberti 阿尔贝蒂*

* Leon Battista Alberti，莱昂·巴蒂斯塔·阿尔贝蒂（1404—1472），文艺复兴时期意大利建筑师、建筑理论家、作家、诗人、哲学家、密码学家，是当时的一位通才。

San Francesco, Rimini 圣弗朗切斯科教堂*，里米尼

* 该教堂为马拉泰斯塔圣殿的前身。1447年由阿尔贝蒂负责改造。

FACADE

AFTEER THE MEDAL BY de'PASTI 仿奠基纪念章*

(VENTURI) (MUNTZ)

* 作者为马泰奥·帕斯蒂，1450年，原件现存柏林国立普鲁士文化博物馆。

（芒茨）

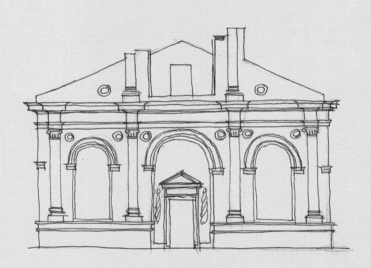

FACADE

S. FRANCESCO AT RIMINI

建于里米尼的圣弗朗切斯科教堂正立面图

(DENKMÄLER DER KUNST)

（摘自《艺术奇迹》）

Santa Maria Novella, Florence 新圣马利亚教堂，佛罗伦萨

SA. MARIA NOVELLA. 新圣马利亚教堂

FLORENCE. 佛罗伦萨

Basilica of Sant'Andrea, Mantua 圣安德烈教堂，曼图亚

PART OF
LONGITUDINAL SECTION
(SIMSON)
纵切面图局部
（辛普森）

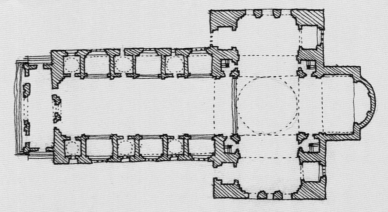

PLAN — (SIMSON)
平面图 — （辛普森）

FAÇADE (MUNTZ)
正立面图 （芒茨）

S. ANDREA, MANTUA
圣安德烈教堂，曼图亚

FACADE OF PORCH

ST. ANDREA AT MANTUA. 圣安德烈教堂门廊正立面图，曼图亚

INSPIRED BY ROMAN TRIUMPHAL ARCH

受到罗马凯旋门的启发

ALBERTI - ARCHITECT.

阿尔贝蒂（建筑师）

— FLETCHER: A HISTORY OF ARCHITECTURE.

摘自《弗莱彻建筑史》

Window in Palazzo Strozzi, Florence 斯特罗兹宫窗户，佛罗伦萨

Life of
DONATO DI ANGELO BRAMANTE

<u>Dates</u>　　born 1444 near Urbino.
　　　　　　died 1514

<u>Parantage</u>

　　Belonged to family of Bramante who held small possessions in the Villas of Monte Asdrualdo, Pistrino and Monte Brandi in the environs of the town of Fermignano, 3 miles from Urbino.

<u>Early Life</u>

(Muntz)　Little imformation.　Studied painting before architecture. (Vasari gives for his master the painter-architect Fra Carnevale of Urbino; probably knew Pietro della Francesca, and perhaps Alberti and Mantegna). His early education gegleted; hardly learned to read and write - some biographers claims that he was unable to write since no autograph had been found. His cleverness made it possible to conceal his shortcomings.

　　It is difficult to believe Bramante had not visited Tuscany.　Some of his works strongly recalled the Florentine monuments.

　　About 1472 or later in 1474 established himsef in Milan, faithful to Sforza until fall of Ludovico il Moro.

　　Some of his paintigs left over show him to have belonged to the Umbrian-Florentine School

布拉曼特生平

日期　　生于 1444 年,乌尔比诺附近。
　　　　　卒于 1514 年

家庭出身

出生于布拉曼特家族。该家族在距离乌尔比诺三英里(约 4828 米)远的费尔米尼亚诺附近拥有蒙特阿斯杜拉多、皮斯特里诺和蒙特布兰地三座别墅的房产。

早年生平

(芒茨)相关记叙甚少。他在从事建筑设计之前曾学过绘画。(瓦萨里提出他的老师是乌尔比诺的画家、建筑师弗拉·卡尔内瓦莱;他可能还认识皮耶罗·德拉·弗朗西斯卡*,或许还认识阿尔贝蒂和曼特尼亚*。)

* Pietro della Francesca,皮耶罗·德拉·弗朗西斯卡(约 1415—1492),意大利翁布里亚画派画家。
* Andrea Mantegna,安德烈亚·曼特尼亚(1431?—1506),意大利画家。

由于早期教育的缺失,他几乎没有学过读写。因为没有发现任何留存下来的手迹,一些传记家称他不会写字。然而其聪明才智足以掩盖缺点。

很难相信布拉曼特没有去过托斯卡纳。他的一些作品很容易让人联想到佛罗伦萨的历史遗迹。

布拉曼特约于 1472 年(或 1474 年)在米兰定居,效忠于斯福扎*,一直到鲁多维科·伊尔·莫罗*政权的覆灭。

* Sforza,斯福扎家族,文艺复兴时期,意大利的贵族家庭。
* Ludovico il Moro,鲁多维科·伊尔·莫罗(1452—1508),米兰国王,斯福扎家族成员,绰号"摩尔人",意大利文艺复兴时期最杰出的王公之一。

留存的一些画作表明,布拉曼特属于翁布里亚—佛罗伦萨画派。

Characteristics

(Muntz)

It is difficult to believe Bramante had not visited Tuscany. Some of his work strongly recalls the Florentine monuments. Interiors of S. M. della Grazia suggests a previous study of the sacristy of S. Lorenzo. The half domes on the exterior of the same church suggest the same treatment at the Duomo of Florence. The arcades of the Canonica of S. Ambrogio sugg suggest the interior treatment of S Lorenzo. The ground floor of the Palazzo S. Blaise at Rome recalls the rustication of the Pitti Palace.

Gey muller says of th influence of th architecture of Lombardy on Bramante:

"The disposition of the piers and vaults as they were specially developed in Lombardy must have had a great influence on Bramante. One must admit too that the Ducal residences at Pavia, Vigevano and of Milan, so imposing in aspect, awakened in him inspirations which he would have sought in vain anywhere else in Italy. The cathedral and the chuch of S. Lorenzo at M Milan also exerted their influene. Sure of himself he did not hesitate to combine a lombard-romanesque archivolt with the new forms in the door on the south side of th Cathedral at Como in 1491."

建筑特点

（芒茨）

很难相信布拉曼特没有造访过托斯卡纳。他的一些作品让人立刻想起佛罗伦萨的历史遗迹。从圣马利亚感恩教堂的内部推测，此前他曾研究过圣洛伦佐的圣器收藏室。而这座教堂外部的半圆形穹顶的设计与佛罗伦萨大教堂的穹顶类似。圣安布罗乔教堂的拱廊与圣洛伦佐教堂室内的风格相似。罗马圣布莱斯教堂的首层让人想起皮蒂府邸粗犷的石材砌造手法。

盖米勒*曾谈到伦巴第建筑风格对布拉曼特的影响。

*Heinrich von Geymüller，亨里希·冯·盖米勒（1839—1909），建筑史学家，主要从事文艺复兴时期建筑的研究。

"伦巴第建筑对基柱和穹顶的布局一定极大地影响了布拉曼特。同样必须承认的是，帕维亚、维杰瓦诺和米兰的总督府邸都非常壮观，启发了他的灵感，这是他无法在意大利的其他地方获得的。米兰的大教堂和圣洛伦佐教堂也对他产生了一些影响。在科莫大教堂南立面大门的设计中（1491年）把握十足的布拉曼特果断将伦巴第-罗马式风格的穹窿形装饰和新的形式融合在了一起。"

S. M. DELLE GRAZIE, MILAN
圣马利亚感恩教堂，米兰
(VALERI) （瓦莱里）

Santa Maria delle Grazie, Milan 圣马利亚感恩教堂，米兰

S.M. della Grazia, Milan

1492 -(Anderson)-

1490 -(Simpson)-

By Bramante.

(Simpson)

To the Gothic nave of S. M. Della Grazia, Bramante added the well-known east end,

<u>Plan</u>   Consisting of a central square with semicircular cpases [apses] to the North and Sount, and a long chancel eastward.

The walls are thin, and, the strong corners notwithstanding, it is doubtful if they could have supported a masonary dome.   Whether he intended one or not is uncertainl   The sixteen sided lantern with its flat-pitched roof, which covers and protects the inside dome, is quite in accordance with nothern traditions; but it was added by others after he left Milan. Each side of the upper story is pierced by a cirdular window; and these 16 windows correspond with a equal number of openings in the inner dome. Fascinating as the exterior is, in its mixture of marble, terracotta, and brick, it is not a good influence.   The simpliciy of the rectagular windows and recesses in the lower portion is delightful, and shows that the architectural side of Bramante's mind was awakening; but beautiful through some of the detail is above, the storéys do not combine to form a happy composition.   In the interior is seen the favarite device of the architect's - two concentric archivolts, at same distance from each other, united by a series of circles, which fill the space between them.

圣马利亚感恩教堂,米兰

1492 年(安德森)

1490 年(辛普森)

布拉曼特的作品

(辛普森)

在圣马利亚感恩教堂哥特式的中堂东侧,布拉曼特增建了著名的半圆形后殿。

方案 (后殿)中部为广场,南北为半圆室,东向为长圣坛。

尽管拐角很坚实,墙却很薄,因此不太可能承受住砖石圆屋顶的重量,至于布拉曼特当时是否想要修建这样(砖石)的穹顶,现在已不得而知。顶塔呈十六边形,平坡屋顶,覆于穹顶之上,非常符合北方的传统;但这是在他离开米兰以后,其他人扩建的。上层的每一面都有一扇圆形窗,十六扇窗和内部的穹隆的十六处开口相对应。尽管其混合了大理石、赤陶和砖的外部结构非常迷人,但这并没有产生什么好的影响。下层简洁的矩形窗和壁龛让人赏心悦目,这也表明布拉曼特的建筑思想已开始走向成熟。尽管以上的一些细节十分精美,楼层之间的结合却并不令人满意。内部可以看到建筑师最钟爱的手法——两个同心的拱门饰,两者之间有一定的距离,由一系列的圆连接,填充了它们之间的空间。

(R.I.B.A.)

Lower half only of Renaissance part was executed under direction of Bramante. Part of decoration in marble instead of terra cotta. The sacristy is decorated on vaulted ceiling with knoted cord or braid called in Milanese "Gruppi". In Urbino a drawing for S. M. della Grazie showing corrections by Bramante and a form of lantern suggesting Tempietto. This shows that in 1492 he was finiliar with classic forms.

(Geymuller)

The east end of S. M. della Grazzia begyn by March 29, 1492. Geymuller thinks that in the upper part if Bramante's drawings were followed at all they were followed darelessly. Count Pompeo Gherardi who died recently in Urbino had a drawing of S.M. della Grazia either by Bramante or a copyist in which the proportions were much finer. Geymuller thinks the cloister and sacristy are by Bramante, but is uncertain regarding doorway.

Geymuller thinks it possible that the architecture shown in the crucifision, a fresco by Donato Montorfano in the refectory may have been drawn by Bramante. Bossi attributes S.M. della Grazia to Leonardo de Vinci. No document to prove this and it is not mentioned by Fra Luca Paccioli who dedicated his manuscript to Ludovico il Moro in which he speaks of the last supper but no reference made to Leonardo as architect of church.

（英国皇家建筑师学会*）

下半部分中只有文艺复兴风格的部分是在布拉曼特的指挥下建造完成的。部分装饰以大理石取代了陶瓷砖。圣器室的顶棚饰有节绳和穗带，米兰语称其为Gruppi。教堂所在地乌尔比诺有幅图纸展示了布拉曼特对教堂所作的修正，其中穹窿顶塔样式使人联想到坦比哀多礼拜堂。这表明他在1492年就已熟悉了古典样式。

* R. I. B. A., Royal Institute of British Architects, 英国皇家建筑师学会，于1834年以英国建筑师学会的名称成立，1837年取得英国皇家学会资格。与美国建筑师学会（AIA）并称当前世界范围内最具知名度的两大建筑师学会。

（盖米勒）

圣马利亚感恩教堂半圆形后殿的修建始于1492年3月29日。盖米勒认为，就上半部分而言，即便当时真采纳了布拉曼特所绘的图纸，施工者的工作也不够认真。不久前在乌尔比诺去世的蓬佩奥·盖拉尔迪伯爵有一幅圣马利亚感恩教堂的手绘图，推测由布拉曼特绘制（抑或是摹本），其比例更为精细。盖米勒认为回廊和圣器室都是布拉曼特的手笔，至于门廊则不能确定。

盖米勒认为，由多纳托·蒙托尔法诺*绘制的修道院餐厅中的壁画《耶稣受难》中的建筑有可能是布拉曼特所画。博西*则认为圣马利亚感恩教堂为达·芬奇所设计，不过这一观点并无文献支持。此外，卢卡·帕乔列*在他献予鲁多维科·伊尔·莫罗的手稿中虽提到了《最后的晚餐》，却并未涉及达·芬奇担任教堂建筑师一事。

* Donato Montorfano, 全名Giovanni Donato da Montorfano, 乔瓦尼·多纳托·蒙托尔法诺（约1460—1502或1503），意大利画家，代表作品《耶稣受难》。
* Bossi, 全名Giuseppe Bossi, 朱塞佩·博西（1777—1815），意大利画家，伦巴第大区新古典主义文化的代表人物。
* Fra Luca Pacioli, 卢卡·帕乔列，又译帕西奥里（1445—1517），现代会计之父，亦在艺术方面有很深的造诣。

S. M. DELLE GRAZIE, MILAN
圣马利亚感恩教堂，米兰．

—(SIMPSON)—
—（辛普森）—

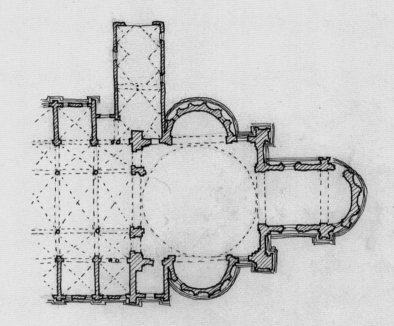

S.M DELLE GRAZIE, MILAN
圣马利亚感恩教堂，米兰
EAST END 半圆形后殿

—(SIMPSON)
—（辛普森）

PALAZZO CANCELLERIA

(Letarouilly)

One of the most beautiful and magnificent palaces in Rome. Situated between Piazza Navona and Piazza Farnise. Since Clement VII, has served as habitation of the Cardinal vice Chancellor and residence of the Apostolic Chancellor xxx. In 1435, Cardinal Ludovico Mezzarota of Padua made some repairs but, after the inscription engraved upon the frieze of the first floor, th Cardinal Raphael Piareo, nephew of Sixtus IV and enriched by him, rebuilt at his expense a part of the foundations. Bramante was in charge of this construction.

(Vasari says Bramante came to Rome in 1500, but an insctiption on the Cancelleria; which is incontestably Bramante's work, bears the date 1495.)

The major part of this work is constructed of stones from the Coleseum, the arch of Gordien and the antique baths existing in the villa Ceretta situated to the east of the thermes of Diacletien.

In plan three difficulties to meet:

1. To confine his construction to the irregular perimeter of the plot.
2. To build a church upon ground already concecrated, with the probable obligation of utilizing old walls.
3. Procuring a number of ofices for the use of the apostolic chamber, and giving to the building a nonumental character inspite of the small divisions called for by such a program: now all these conditions are fulfilled without one seeing the efforts they have cost. The entrance to the palace is commodious, the court of a simple ordinance, unoble and regular, the doors spacious, the stairway prominent, and the communication with the ch urch easy and direct. The little court with its entrance upon a lateral facade,

坎榭列利亚宫

(勒塔鲁伊)

坎榭列利亚宫坐落于纳沃纳广场和法尔尼斯广场之间，是罗马最宏伟壮丽的宫殿之一。自克雷芒七世起，这座宫殿一直作为教区秘书长的住所和教皇文书院。1435年，帕多瓦的红衣主教鲁多维科·梅扎罗塔[*]对宫殿进行了一些修整，然而在他把铭文刻在第一层的檐壁上之后，红衣主教拉斐尔·里阿里奥（教皇西克斯图斯四世之侄，因其叔而变得富有）出资对一部分宫殿进行了重建。布拉曼特负责主持这一重建工作。

（据瓦萨里记载，布拉曼特是1500年来到罗马的，但坎榭列利亚宫的一处铭文的刻文时间为1495年，而该宫确出自布拉曼特之手。）

宫殿主体的石料分别取自古罗马竞技场、戈尔迪安宫的拱门以及戴克里先浴场东面Ceretta庄园的古老浴池。

设计中的三大难题：

1. 宫殿须建在一块不规则的划地上；
2. 教堂的建址已被圣化，因此很可能须延用原有墙体；
3. 宫殿要依照教皇需求安排若干房间供教皇作会议厅用，同时，尽管要设计出这些封闭的小房间，整个建筑还一定要体现出它的神圣宏伟。现在人们看到设计出的宫殿鬼斧神工般地满足了所有的要求：入口宽敞，中庭优雅简洁，内门宽大气派，楼梯位置明显，且与整栋建筑连接得自然、直接。在宫殿侧面设有入口的小庭与中庭和花园共同形成了实用的通路，满足了多种需求。

[*] Ludovico Scarampi Mezzarota，鲁多维科·斯卡兰皮·梅扎罗塔（1401—1465），又名Ludovico Trevisano，鲁多维科·特雷维萨诺，生于帕多瓦，意大利红衣主教。

obtains at the same time, with the great court and the garden, a useful communication and thus satisfies a number of needs.

The facade which is sensible composition and of a noble character, shows a one time beautiful lines, happy divisions and a number of rich details. It is imprinted to the highest degree with the manner of Bramante and should be considered as the true type of it.

It is to be regretted that Vignola and Dominico Fontana, who did the works on this palace, did not get the details more in harmony with he ensemble. The projecting mass at the ends of the facade are at this time rare but which has been used frequently since. They give more strength to the angles. Their projection though small is sufficient to break the too long drawn out lines of the facade.

Facade entirely construced of travertine except for the windows of first floor and columns of the entrance door which are of marble.

The mouldings are delicate and fine in pofile. The cornice is repeated in the court and the composite caps are repeated in other works of Bramante, notable the Palzzo Girauad and Doria. They seem inspired from antique caps ad harmonized perfectly with the elegance of the windows.

Nothing more elegant than the detail of these beautiful windows, they are of white marble and executed with great care. The arabesques with which the pilasters are decorated are of a fineness and variety quite enchanting. The balcony is remarkable for the richness of its sculptures and show his good taste and judgement in the happy combination of consols with panels and the grace of the ingenious arabesques.

The architecture of te court is noble, elegant and picturesque. The effects of perspective have been calculated here in a brilliant manner.

建筑的正面构图合理，显示出高贵的气质——线条诉说着曾经的优美，分割恰到好处，还有丰富的细节装饰。这座宫殿被认为是布拉曼特设计风格的极致体现，是他的代表作之一。

遗憾的是，负责具体建造工作的维尼奥拉*和多梅尼科·丰塔纳*并未能使宫殿的细节设计很好地与整体风格统一在一起。外立面边缘处的突出部分增强了墙角的效果，这种设计在那个时代还很少出现，但后来则得到了大量应用。突出的长度虽然不大，却足以淡化外围的过度拉长之感。

除了第一层的窗结构和入口的支柱为大理石质外，整个外立面均采用了石灰华。

模塑装饰精致且轮廓清晰。中庭中同样采用了檐口的设计，混合式柱头也在布拉曼特的其他作品中有大量的应用，尤其是托洛尼亚宫与多利亚画廊。这些柱头的设计似乎是受了古典柱头设计的启发，与优雅的窗户完美地融合在一起。

窗户上的精美装饰优雅绝伦，为白色大理石质，雕工甚为细致。窗旁的壁柱雕以阿拉伯花饰图案，其细腻多样，使人着迷。阳台设计了丰富的雕饰，带有镶板的悬臂与优雅独特的阿拉伯花饰图案搭配在一起展现出建筑师独到卓越的建筑品味。

中庭的设计高贵优雅，精美如画。透视效果在这里得到了完美的运用。

* Vignola，全名Giacomo Barozzi da Vignola，维尼奥拉（1507—1573），意大利建筑师，16世纪欧洲手法主义的代表人物。
* Domenico Fontana，多梅尼科·丰塔纳（1543—1607），文艺复兴时期欧洲工程师。

The old columns from the basilica of S. Lorenzo were used to fashion a two story portico which procured a brilliant effect, but the work was far from been finished. There were a number of needs yet to be satisfied. It was necessary to establish offices, creat living apartments, build new floors and as they lacked all kinds of materials, they had to use travertine to replace marble for upper stories. The pilasters placed above th columns in continuing these vertical lines and in order to give the motive a better feeling a wall of good bricks served to separate them, a beautiful combination of materials which shows all the resources and all the fertility of genius, which is on occassions to overcome difficulties which he had to conquor in spite of the poor means of execution.

To strengthen the corners, the architect prefered piers instead of columns. This offers the advantage not only of satisfying the eye and showing the corners of the court better but also allows the arches to finish without bumping each other.

(Simpson)

The earliest building in Rome atributed to him. It was begun about 1495, before Bramante resided in Rome, but he may have prepared the design during his visit there in 1493. Although beautiful in detail it is poor in scale in comparison with his later work. This defect is accounted for if the design were made before he came strongly under the influence of the old remaind in the capital. The facade is misleading, as no indidation is given that behind it there are two buildings, one being a church, S. Lorenzoin Damazo, to which the smaller door gives acces. To some extent it recalls Alberti's Palazzo Rucellai, Florence. But the spacing of the pilasters and the windows are different. The latter are all founded on th windows of the

宫殿双层门廊的支柱运用了圣洛伦佐教堂（门柱）的古老风格，效果斐然。但整个工程还远没有完成，还有很多需求呕待解决：需要修建办公室、卧室和新的楼层。鉴于当时材料的匮乏，他们不得不采用石灰华代替大理石建造上面的楼层。支柱上方设计了壁柱，很好地将这些垂直线条贯穿了起来；为了使这种搭配更加协调，建筑师又在两者之间加了一条精致的砖带。对材料恰到好处的组合运用展现了建筑师的智慧和他丰富的设计天赋。尽管施工的条件很差，建筑师仍然克服了建造过程中出现的各种困难。

为了使庭院的角落处更加稳固，建筑师独具匠心地在中庭拐角处采用了方柱取代圆形石柱，这样不仅视觉上感觉更流畅，拐角更明晰，而且又避免了拐角衔接处的两个拱门撞到一起。

（辛普森）

该建筑被认为是布拉曼特在罗马最早的作品。尽管这座宫殿始建于1495年前后，即在布拉曼特还没有来罗马定居之前，但是他可能在1493年来罗马访问时就开始准备对坎榭列利亚宫的设计。虽然建筑细节精美，但其与布拉曼特后期的作品相比则缺少些许层次感。如果这个宫殿的设计是在他来到罗马定居之前，也就是还没有受到罗马建筑古迹设计的强烈影响之前完成的，这个缺陷就不难理解了。宫殿从正立面看很具有迷惑性，完全看不出其后有两座建筑，有一个是圣洛伦佐教堂，可从（右侧）小门进入。这样的设计多少让人想起了由阿尔贝蒂在佛罗伦萨设计的鲁切拉宫，但是壁柱间的间隔和窗户的设计与之不同。

Porta de' Borsari, Verona, which they closely resemble, although the detail is more refined and delicate. Bramante knew Verona well, and this gatway was probably one of his chief sources of inspiration before he came to Rome. The grouping of the Pilasters in pairs, with walling between, is particularly happy, the narrow spaces framed by each pair being half the width of the wider bays in which are the windows. This treatment was a favorite of his; and that fact, coupled with the design of the windows, affords presumptive proof that he was the architect of the building. The two tiers of pilasters are approximately equal in height, but as the upper tier runs through two storeys; what might otherwise have been disagreeable proportions are avoided. The pilasters on the back and side elevations are equally spaced, and not in pairs. The walling of these portions is brick; only the front is stone throughout.

(Muntz)

The Cancelleria, or the palace of S. Domasso, built for the use of the Cardinal Raphael Riario, nephew of Sistus IV. One of his first work in Rome and the one in which he passed from his Lombard to his classic manner. He breaks absulutely with the style of S.M. della Grazie and S. Satiro and henceforth the least possible ornament, and no cnadelabra shafts.

The facade, like Alberti's Rucellai, employs a fine rustication together with pilasters, but an entirely new arrangement. The ground floor has no orament except the arched windows, the pilasters are reserved for the first ad second story. The alternate spacing of the pilasters has been spoken of. Letarouilly points out the delicacy ad the strength of the profiles, and the fine combination and relation of the mouldings. Nothig is more beautiful than the details of the windows, particularly the

后者可以看出与维罗纳的博尔萨里门上的窗户非常相似，但在细节处理方面更精美细致。布拉曼特对维罗纳的建筑有很深的研究，在来罗马之前，这个门道或许是他设计的重要灵感来源之一。正立面的壁柱成对排列，每对间以墙体隔开，比例恰到好处，他把成对壁柱间的距离缩小到含窗开间宽度的一半。这种处理手法是布拉曼特所钟爱的，再加上窗户的典型设计，我们可以据此断定这座宫殿是出自于他之手。两排壁柱的高度几乎相等，但由于上排壁柱贯通两层，不协调的比例便得以避免。建筑背面和侧立面的壁柱是等距分布的，也并未成对出现，墙体由砖砌成，只有正立面是完全用石料筑成的。

（芒茨）

坎榭列利亚宫，亦称圣达马索宫，是为西克斯图斯四世的侄子、红衣主教拉斐尔·里阿里奥修建。它是布拉曼特在罗马的第一个建筑作品，也使他完成了从伦巴第风格向古典风格的转变。他彻底脱离了圣马利亚感恩教堂和圣沙弟乐圣母堂的建筑风格，装饰能简则简，同时摒弃了枝状大烛台的使用。

像阿尔贝蒂的鲁切拉宫一样，宫殿的正立面以粗面石和壁柱相搭配，但是设计方式则完全不同。一层除了拱形窗，没有其他装饰，二层和三层则都设计了壁柱，它们之间的间隔变化上文已经述及。勒塔鲁伊指出，整座宫殿轮廓精致明晰，嵌条的搭配组合细腻巧妙。窗户优雅精巧的细节设计无与伦比，尤其是窗框壁柱上的阿拉伯式花饰雕刻。

arabesques on the pilasters.

In the court of the palace Bramante has contras with the two lower stories, with their open arcades, the upper story relieved only by pilasters. Bramante had at his disposition th columns of the ancient basilica of San Lorenzo in Damaso demolish ed at eh time of the construction of the palace, and which originally came, it is said from the Portico of Pompey. This arrangement alone procured a brillian t effect but it was necessary to have more stories. The was no more of this old ready made material. Bramante turned to travertine to replace the marble in the upper stories, pilasters placed above the columns continue the vertical lines, and in order that these shall count for more they detach themselves from a brick wal.

The Rose of the Riario is lika dominant note in ornamntation and is used ont eh capitals of the columns, on the corner piers, in the spandrels of the arcades. Perhaps this idea is carried too far.

(Geymuller)

Great puzzle as to date. Could be by no one other than Bramante. Bramante came to Rome im 1499. The facade has inscription tat it was begun in 1495. In a letter from a Milanese to the Roman architect Andrea Vici, Bramante's absence from Milan in December 1493 is noted. Thought tat Cardinal Raphael Riario may have heard of Bramante in his native town Savone. Bramante speaks in a sonnet of a visit to savone.

The facades with rustication and he orders was known in Milan inpainted arch. Casa Silvestri, atrributed to Bramante ÷ 1475.

The design of the door found by Letarouilly and claimed by him to be

宫殿中庭下两层设有开敞的拱廊，最上层仅用壁柱突显，形成对比。布拉曼特还用上了建筑宫殿时从古老的圣洛伦佐教堂拆下来的支柱，据说这些柱子可以追溯到庞培门廊。仅仅是这个设计就产生了异乎寻常的效果，但楼层还需继续向上修建。此时，原有的成料早已用完，布拉曼特就在上层的建筑中用石灰华来代替大理石，在支柱上方建造壁柱以延伸垂直的线条，产生一种壁柱独立于砖墙存在的奇妙效果。

里阿里奥家族的玫瑰图案是这个设计中的核心装饰元素，出现在支柱的柱头上、拐角的方柱上、拱廊的肩拱上。这个元素恐怕用得过于频繁了。

（盖米勒）

有一个大的疑问是关于建筑的日期。宫殿除布拉曼特外不大可能出自他人之手。布拉曼特1499年来到罗马，但是宫殿正立面的铭文上却写着建筑动工的日期是1495年。一封米兰人写给罗马建筑师安德烈·维西的信中提到，1495年12月布拉曼特不在米兰。红衣主教拉斐尔·里阿里奥可能在他的家乡萨沃纳就听说过布拉曼特。布拉曼特的一首十四行诗也记叙了他浏览萨沃纳的经历。

坎榭列利亚宫粗面石的外观和柱式特点曾见于米兰的彩绘拱门上。建于1475年的丰塔纳·席维斯特瑞之楼被认定为布拉曼特的作品。

据记载，勒塔鲁伊发现了大门的设计图并声称是出自布拉曼特之手，盖米勒则认为不管是线条风格还是笔迹都表明是安东尼奥·达·圣加罗的作品。

Bramante's.   Geymuller says both in line and writing is Antonio San Gallo.

(Geymuller)

From the List.

Facade executed from design of Bramante 1495.

Court under personal direction of Bramante 1500-1514.

Drawing for great door attributed to him is by Sangallo

(Gumaer)

The projection at end is significant at this early period.   The rythmic spacing is to credited to Alberti.   It is quite different from coupled coled columns.

In court, when he comes to corner he used piers.   It satisfies the eye.   Small entrance by Vignola.

（盖米勒）

见作品表。
宫殿正立面依据布拉曼特 1495 年的设计建造；
中庭则在布拉曼特亲自主持下于 1500 – 1514 年间完成。
归入其名下的大门的设计图应归为圣加罗。

（古米尔）

建筑拐角处的突出式设计在早期具有十分重要的意义，富于节奏的间隔设计应归功于阿尔贝蒂的智慧，这与成对圆柱的样式截然不同。

中庭拐角处恰当地应用方柱来代替圆柱，给人以流畅的视觉效果。小型入口由维尼奥拉完成。

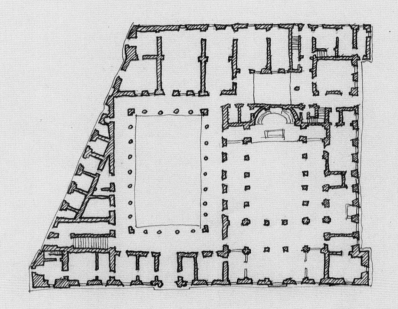

PALAZZO CANCELLERIA.
坎榭列利亚宫

—(LETAROUILLY)—
—（勒塔鲁伊*）—

*Letarouilly, 勒塔鲁伊（1795–1855），法国建筑史学家，绘制了大量罗马古建筑图纸。

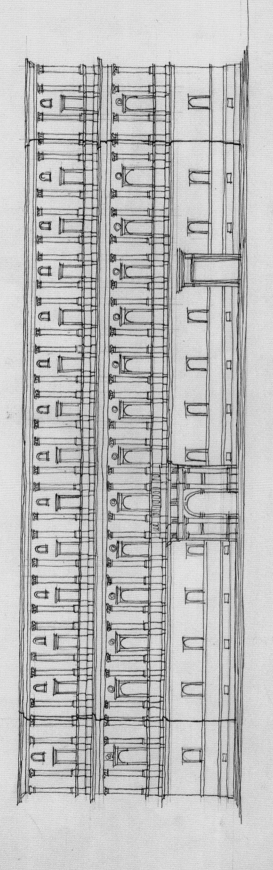

PALAZZO CANCELLERIA ROME
坎谢列利亚宫 罗马

-(LETAROUILLY)-
——(勒塔鲁伊)——

PALAZZO CANCELLERIA ROME
坎榭列利亚宫 罗马

COURT
中庭

—(LETAROUILLY)—
—（勒塔鲁伊）—

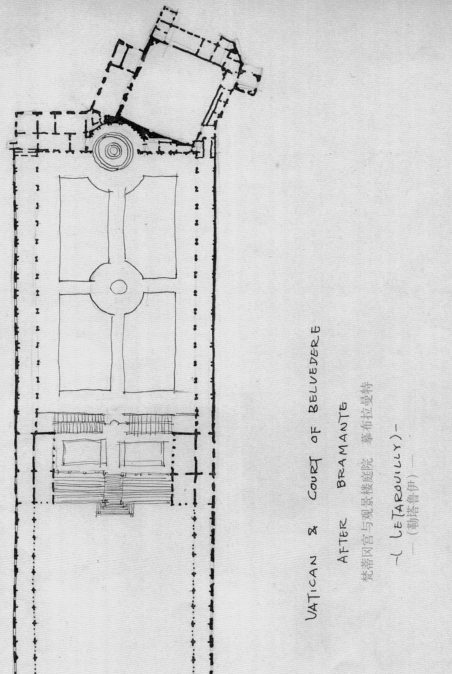

Vatican & Court of Belvedere
After Bramante
梵蒂冈宫与观景楼庭院 掌布拉曼特
—（LETAROUILLY）—
—（勒塔鲁伊）—

Vatican 梵蒂冈宫

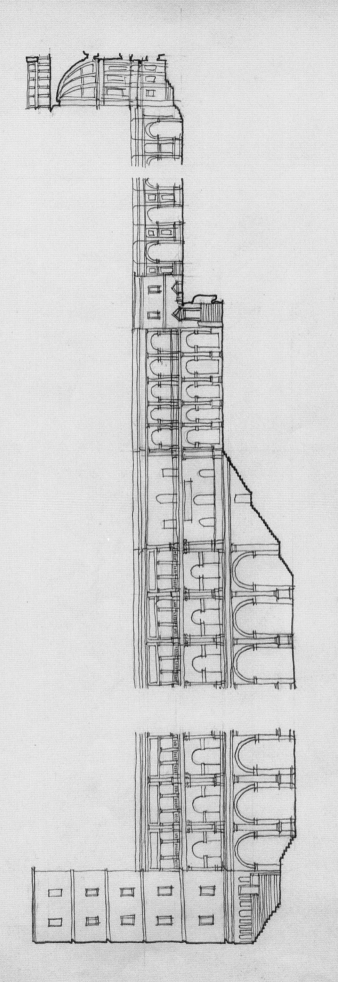

COURT OF BELVEDERE
RESTAURATION OF BRAMANTE'S PROJET
观景楼庭院 布拉曼特工程复原图
—(LETAROUILLY)—
—(勒塔鲁伊)—

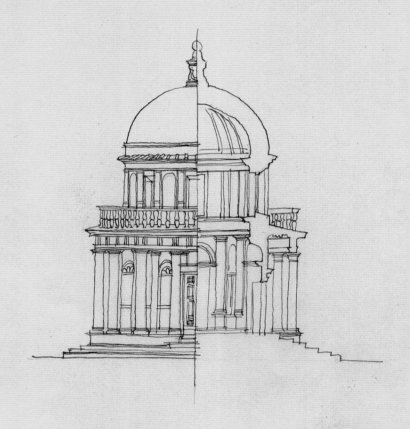

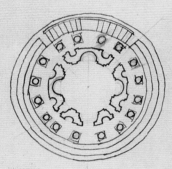

IL TEMPIETTO ROME
坦比哀多礼拜堂，罗马

— (GROMORT) —
—（格罗莫尔）—

IL Tempietto, Rome 坦比哀多礼拜堂，罗马

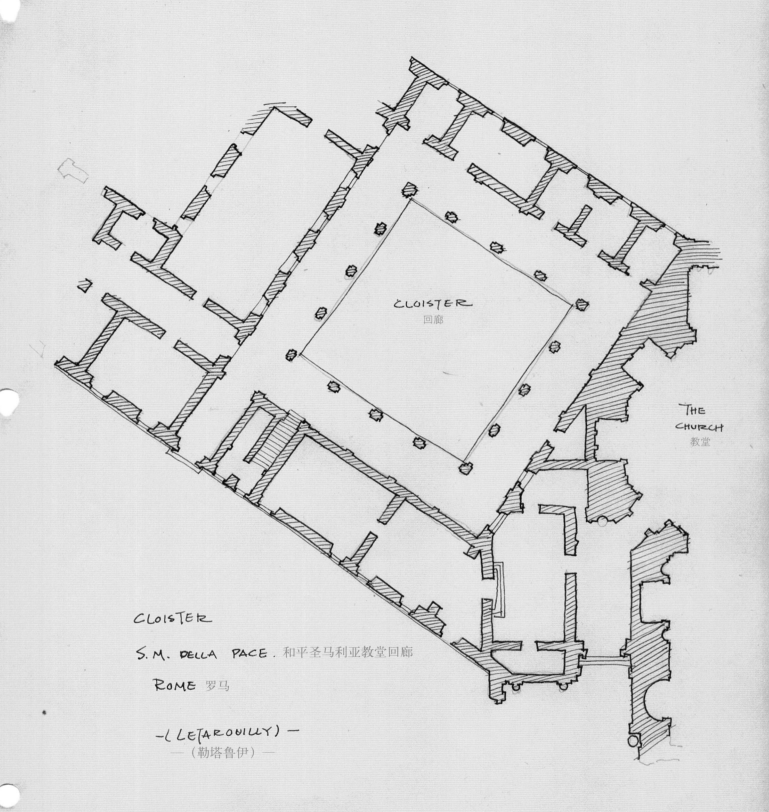

Santa Maria della Pace, Rome 和平圣马利亚教堂，罗马

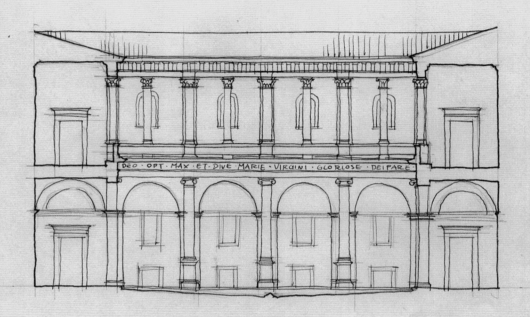

CLOISTER
S.M. DELLA PACE. 和平圣马利亚教堂回廊
ROME 罗马

—(LETAROUILLY)—
—（勒塔鲁伊）—

St. Peter's Basilica, Rome 圣彼得大教堂，罗马

0 12 24 36 48 60 72 84 96 M.

BRAMANTE'S SCHEME FOR
ST. PETER'S.

布拉曼特关于圣彼得大教堂的设计方案

ST PETER'S

(R.I.B.A.)

The beginnig of the construction of S. Peter's goes back to the first and most ardent champion of th Renaissance, Nicholas V. Almost as soon as his eletion 1447, he ordered th demolition of the shops which encumbered the atrium. As soon as he was in possession of the great sumns furnished by the Jubilee of 1450 he started the reconstruction of the Basilica of Constantine. The death of Nicholas V. 1455 interupted the Work. The work of Alberti and Rossellino has howerer more than an historic interest. Modern writers are agred that the foundations of the 2 Florentines were utilized and that their plan was taken into accou nt up to the beginning of the XVI century.

It is generally believed that Julius II was the first to continue the work. This is an error. 35 years before Julius II came to the throne Paul II, too long calumniated, began work on the tribune.

It was reserved, however, for Julius II. to push th work to such a point that the completion was only a matter of time an money. Before arriving at the idea of a complete redonstruction, his plans passed through diverse stages. His first plan was to complete the apse begun by Nicholas V and to place there his mausoleum. The execution of this work had been conflided to M. Angelo since March 1505. The architects at first enterd into his views but plans more and more brilliant were developed. The final result was the withdrawal of M. Angelo and the adoption of a plan for the entire reconstruction of the basilica in the Renaissance style. Jan. 6, 1506 Julius wrote to the King of England to announce this news and solicitd his cooperation. On the 18th of the following April the first stone was laid.

圣彼得大教堂

（英国皇家建筑师学会）

圣彼得大教堂的建造始于文艺复兴第一个也是最为热情的倡导者，尼古拉五世。1447年他登基不久就下令拆除所有阻碍中庭的店铺。到1450大赦年，尼古拉五世获得了充足的资金，下令开始重建君士坦丁教堂。1455年尼古拉五世去世后工程中断。阿尔贝蒂和罗塞利诺*的工作并不只具有历史价值。现代学者一致认为这两位佛罗伦萨人的设计为以后的建筑设计奠定了基础，他们的方案直到16世纪早期都仍然受到重视。

人们大都认为尤里乌斯二世是第一个继续主持这项工程的人，这种说法是错误的。早在尤里乌斯二世继位之前35年，保罗二世就已经开始了教堂的施工。

但工程的主推手最终还是尤里乌斯二世。在他的推动下，工程的完成只是时间和金钱的问题了。在决定推倒重建之前，尤里乌斯二世的设计和想法经过了许多非常不同的阶段。他起初是想完成尼古拉五世时期就开始建造的半圆形殿，并想将其作为自己的陵墓。从1505年3月起，米开朗琪罗受任主持这项工程的施工建设。起初，建筑师们认同他的观点，但后来不断有更精彩的方案提出来。最终米开朗琪罗退出了工程，而教皇则采纳了以文艺复兴风格完全重建教堂的方案。随后，他于1506年1月6日致信英国国王告知他重建教堂的决定，同时寻求英国的合作。同年4月18日，圣彼得大教堂破土动工。

* Bernardo Rossellino，贝尔纳多·罗塞利诺（1409—1464），文艺复兴时期意大利雕塑家和建筑师。

On the 6th of April, 1506, the Pope put 7500 ducats at the dispositon of the architects. Then comes under date of April 20, the constitution of the fidejusseurs. The 23rd of April the money is turned over to the contractors who are called "magistri architecti". These were originally five in number. The contract signed by them shows that each brasse of the wall was to be paid for at a fixed tarif. For th e columns ad the capitals there was a fixed sum determined in advance. For th e other work the workmen were paid by the day. Bramante had charge of all the payments, one finds his name on almost every page of the pontifical accounts. A post so important as architect-in-chief of S. Peters was not obtained without some fighting and perhaps intrigue. Candidates were not lacking, to begin with Giuliano da San Gallo, the old frien of the Pope, and Fra Giocondo.

Processes

Bramante began by destroying half of the old Basilica(this destruction of a shrine venerated above all others gave him the name of "master demolisher").

The work advanced rapidly. We have mentioned the contract concluded in 1506 with the masons charged with constructing the foundations and to build the piers of the cupola. The stone cutters from March 1507 worked on the capitals of the new basilica. On the first of March 1508 the contract was signed by which Francesco di Dominico of Milan, Antonio de Jacopo of Ponte a Siene, and Benedetto di Giovanni Albini of Rome, all three designated as "scarpellini", promised to execute a certain number of capitals on the models of those of the Pantheon.

The work suffered for lack of funds; the Pope did not depend on his regular in come. Numbous letters show that all Europe was called upon to contribute. If we are to believe the author fo the satirical poen "Simia", published at Milan in 1517, the indulgences saved the Pope from dœaeing on his

1506 年 4 月 6 日，教皇拨款 1500 达克特供建筑师使用。到 4 月 20 日那天，组成了担保人团队，23 日钱被交付给了被称为"主建筑师"的承包者们。最初的承包者是五个人，签约合同规定墙上的黄铜器都按照固定的价格付费。对于支柱和柱头，都预先确定了费用。其余工作按日计费。布拉曼特负责支付所有的费用，他的名字几乎出现在教皇账目的每一页中。要夺得像圣彼得大教堂首席建筑师这样重要的职位必定少不了竞争甚至计谋。众多候选人中包括如教皇的老朋友朱利亚诺·达·圣加罗和乔瓦尼·焦孔多。

进程

布拉曼特首先推倒了大半个旧教堂。（他因为推倒了世人最为尊敬的神殿而得名"摧毁大师"。）

工程进展很快。上面提到 *1506* 年签订的合同约定，石匠负责建造圣彼得大教堂的石基，并且将完成圆屋顶的方柱施工。自 *1507* 年 *3* 月，凿石工开始雕刻建造新教堂的柱头。*1508* 年 *3* 月 *1* 日，被任命为石匠的三个人：来自米兰的弗朗西斯·多梅尼科，锡耶纳的安东尼奥·雅各布和来自罗马的贝尼代托·吉奥瓦尼·阿尔比尼，签订合约承诺建造出一定数量以希腊万神庙为范本的柱头。

这项工程一度缺少资金，教皇没有依靠他常规的收入来维持工程的建设。很多信件证明当时教皇号召整个欧洲为圣彼得大教堂的建造捐款。如果我们相信 *1517* 年那首发表于米兰、题为 Simia 的讽刺诗的作者所言非虚，那么教皇则是通过兜售赎罪券而省下了自己的收入。从 *1506* 年 *4* 月到 *1513* 年底的总花销达 *70,633* 达克特。

own income.   The expenditure from April, 1506, to the end of the year 1513 was 70,653 ducats ($706,530).   ~~From published documents it is known that from December 22, 1529 to January 2 of 1543 13 years.~~

Julius II died 1513.   The work continued.   Bramante died a year after.   At his death the four piers which were to support the dome were built up to the level of the cornice.   The arches were constructed and ornamented with caissons, the chapel at the back was almost finished and in many other parts the construction was far advanced.   Vasari says that the cornice on the interior was son fine that it could not be changedwithout spoiling it.   The capitals and all the Doric part of the exterior also showed his talent.   The work could have been completed according to the original conception; but other plans were adpoted ~~which will~~.

(Scheme

Bramante's scheme for S. Peters was a <u>Greek cross with a cupola at the centre</u>.   <u>Four projecting towers formed the angles of the Square</u>, and the <u>four arms of the cross with their ambulatories projected in the form of semicircular tribunes</u>.   <u>Between these and the towers wer porticoes which led under the four small cupolas</u>.   Geymuller says this scheme <u>uited the majesty of the classic buildings with the fantasy of the mediaeval cathedrals</u>.

Br.'s scheme can only be studied in the drawings which are more or less fragmentary, in the medal of caradosso and the altered plan published by Serlio.

Michael Angelo, Bramante's enemy and adversary, when he was finaly consulted , said: "It cannot be denied that Bramante wasthe greatest architect of all times.   It is he who has made the first scheme for S. Peters and

尤里乌斯二世于 1513 年去世。这项工程由布拉曼特继续主持，直到一年后他离世。他去世时，支撑大圆顶的四根方柱已经建到了檐口，拱门已经建造完成并饰以藻井，后面的小礼拜堂也几近完工，其他部分的施工进展也非常快。据瓦萨里记载，内部的檐口设计精美绝伦，已容不得任何改动。外部的柱头和所有陶立克式的部分都展示出布拉曼特在建筑上的天赋神笔。这项工程本可以按照原来的设想修建完成，但是后来又采用了其他的设计方案。

方案

布拉曼特设计的圣彼得大教堂呈纵横等长的十字结构，中心设有一个大圆顶，四个凸出的塔楼建在十字结构正方形的四角处，十字的四臂连同内部的回廊呈半圆形向外突出。四臂和塔楼中间设有柱廊，与十字拐角处的四个小圆顶相接。盖米勒认为这个结构将古典建筑的宏伟和中世纪教堂的神秘幽幻集于一身。

布拉曼特的设计结构只能通过有些支离破碎、不大完整的草图、卡拉多索纪念币以及塞利奥*出版的改动后的设计方案等进行研究。

米开朗琪罗是布拉曼特的对头和竞争对手，在他最终被问及时说："毋庸置疑，布拉曼特是世上最伟大的建筑师。"

* Serlio，全名Sebastiano Serlio，即塞巴斯蒂亚诺·塞利奥（1475–1554），文艺复兴时期意大利建筑师。

this plan is not confused, it is simple, well lighted, well isolated so that it is no way hurts the palace (the Vatican) and its beauty, which is still manifest, has been justly recognized.   Also whoever has digressed from it, as has been done since S. Gallo, has digressed from the truth."

(R.I.B.A.)

  The dome not disigned to stand alone.   There were to be angular towers.

是他最先设计了圣彼得大教堂的结构，他的设计方案没有繁乱之感，简洁大气，透光充足，自成一体，因此不会防碍梵蒂冈宫，其美感至今仍显而易见，所受赞誉可谓实至名归。自圣加罗以来背离此道者，无不是在背离真理。"

（英国皇家建筑师学会）
  除圆屋顶设计外，本来还应该有尖角塔楼。

PLAN OF ST. PETER
ATTRIBUTED TO BRAMANTE (?)
布拉曼特设计（？）的圣彼得大教堂平面图

Comparison of Bramante's Schemes
of St. Peter's

(Simpson)

Bramante's original plan, which was proceeded with, was a square with a projecting apse on each side. Towers were intended at the four corners, and domes over the four chapels on the diagoals round the central dome. The arms were designed to be barrel vaulted as they are now. The four great piers at the crossing were began first, and before Bramante's death in 1514 they were finished, and some of the arches they support turned. All effort apparently was concentrated on this portion, and no attempt made to begin the out side walls. In no other way can the discrepancies in the many tentative plans prepared by Bramante and his immediate successors be accounted for. In all, the central space remains the same, but the external walls vary considerably in outline. No foundations for these can have been built, otherwise the numerous changes could never have been suggested. Bramante either found it exceedingly difficult to make up his mind regarding the exact shape his building should take externally, or else failed to satisfy his clients. In another design attributed to Bramante, although the authorship is uncertain, aisles are carriesd cound the external apses. These are retained, with modification in Peruzzi's plan. (Peruzzi, beforehebwas appointed architect toS. Peter's, acted for some years as assistant to Br.) Inall three plans cdbumns are shown between the great piers. Bramante had probably learnt their scale-giving properties from San Lorenzo, Milan, and his study of ancient buildings in Rome would confirm him in his opinion of their utility.

To Bramante belongs the credit for the plan of the central portion of the

圣彼得大教堂布拉曼特设计结构比较

(辛普森)

  布拉曼特的最初设计（也得到了实行）呈现正方形，每侧延伸出一个半圆室，塔楼分别设在正方形的四个角上，对角线上的四个小礼拜堂环绕中心大圆顶建造，礼拜堂的上方均设有小圆顶。十字形翼部设计有桶形拱顶，正如现在所呈现的模样。十字结构的四个大型方柱是最先动工的，在1514年布拉曼特去世之前就已经完成，并且它们支撑的一些拱门也已经建造完工。主要的精力和设计似乎都集中在了圣彼得大教堂的内部，并未尝试开始教堂的外墙设计。这是唯一能够解释布拉曼特和后来继任的设计师们在诸多底稿中存在出入的原因。总的来说，中央空间的设计没变，但外墙的轮廓图有很大的改动。这部分没有建盖任何基础，否则决不会出现那么多改动建议。要么是布拉曼特实在无法下决心确定这座大教堂究竟采用何种外观，要么是他的设计无法满足雇主的要求。在另一幅被认为是布拉曼特的设计中（虽然作者身份并不确定），外围半圆室为走廊所环绕。这些设计都在佩鲁齐的设计方案中保存了下来（有所改动）。（佩鲁齐在受任为圣彼得大教堂的主设计师之前，曾任布拉曼特的助理多年。）在三个设计方案中，大方柱之间都设有圆支柱。布拉曼特很可能是从米兰的圣洛伦佐教堂设计中学会了具有层次感的布置，同时他对罗马古建筑的研究也应坚定了他制造层次感的想法。

church, and the general internal ordinance, exclusive of the three end bays of the nave.   He made several designs for the interior.   One section shows two orders with a gellery above the aisles, and a quadripatite bault, another a single Doric order, with a gallery carried on columns between the piers as in Roman buildings.   A third shows the design much as executed, a single Corinthian order carrying an entablature from which springs a barrel vault.

圣彼得大教堂中心部分的设计、内部的总体法度（除正殿末端的三个开间外）都应归于布拉曼特。他为内部的建造贡献了很多设计。一张剖面图中的侧廊上部设有双柱型的走廊和一个四分的拱顶。另一个则是单一的陶立克柱式，像古罗马建筑一样，走廊靠方柱之间的圆立柱支撑。第三个则是最接近实际施工的，单一的科林斯立柱，配以古典柱式的顶部，其上建有桶形屋顶。

Bramante's Scheme Compared with S. Lorenzo, Milan

Simpson says that Bramante learnt the scale-giving properties from S. Lorenzo, Milan.

Bramante apparently adopted the scheme of haveing a square plan with semi-circular apses stuck to the four sides.   In the design atfributed to Bramante, the ambulatory is carried around as in the ancient church. Columns are used between piers to give scale.

布拉曼特设计方案与米兰圣洛伦佐教堂结构设计比较

辛普森认为，布拉曼特是从米兰圣洛伦佐教堂的设计中学习了具有层次感的布置特点。

布拉曼特显然采用了（圣洛伦佐教堂的）设计（方式）：方形设计，半圆室向四外突出。在那幅认定为布拉曼特的设计中，回廊也像古教堂一样环于四围，并在（大）方柱间运用圆柱来体现层次。

(Gumaer)

It is criticised that the piers are too week, and some architects believed that it has been changed, this is probably not true.

It is said that they wanted the Latin cross. But more true is that they wanted to cover up the old plan with now plan. The old basilica stood long after the new was commenced, but the old building began tos settle and became dangerous.

Vasari says that it is Bramante's desire to put the dome of Pantheon on the Temple of Peace (Basilica of Constantine). But the statement is not altogether true as the latter has groined vault.

Geymuller's restoration is very fantastic. In projets primitive he does not seem to have shwown much whcih he base upon.

Perruzzi's plan has the best looking poche.

（古米尔）

有人指出四个支撑方柱不够牢固，还有一些建筑师认为这些方柱曾经被更换过，但这些可能并非真实。

据说后来的建筑师们想要建造拉丁（横短竖长）十字架结构，但更确切地说他们是想用新的设计方案来覆盖旧的设计，新教堂动工很久后，旧教堂依旧存在，但是旧建筑已经开始逐渐下陷，成为危房。

瓦萨里说是布拉曼特想要为和平殿（君士坦丁大教堂）设计希腊万神殿式的大圆顶。但是后者为穹窿屋顶，瓦萨里所言并非完全真实。

盖米勒的复原美仑美奂，在最开始的一些项目里，他看起来并没有很多的依据。

佩鲁齐的设计图涂黑部分最为精美。

SECTION AFTER 摹布拉曼特设计的剖面图

BRAMANTE

—(LETAROUILLY)—
—(勒塔鲁伊)—

ELEVATION AFTER 摹布拉曼特设计的立面图

BRAMANTE'S DESIGN

—(LETAROUILLY)—

—（勒塔鲁伊）—

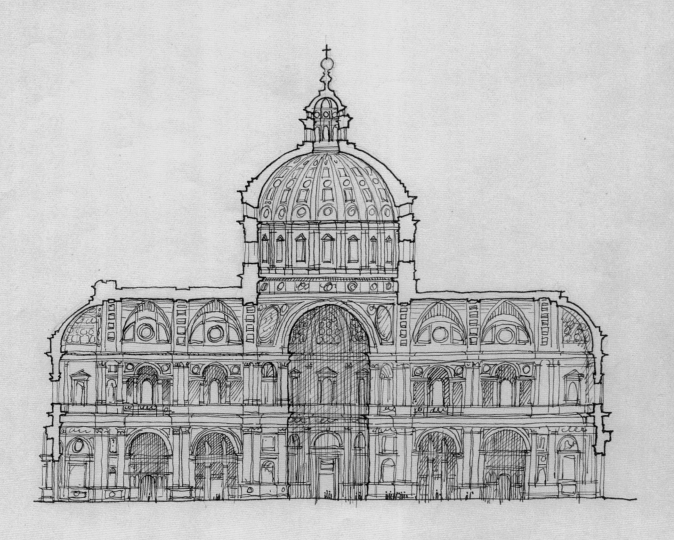

RESTORATION OF ST. PETER'S
AFTER BRAMANTE'S PLAN

BY GEYMÜLLER.

摹布拉曼特关于圣彼得大教堂平面设计图的复原图
盖米勒

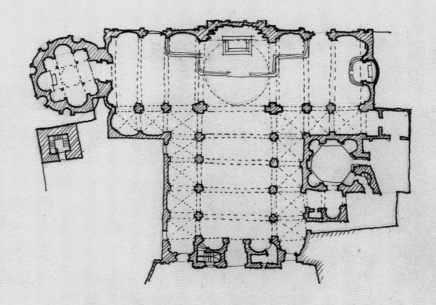

S.M. di SAN SATIRO, MILAN.
圣沙弟乐圣母堂，米兰
(VALERI)
(瓦莱里)

Santa Maria presso San Satiro, Milan 圣沙弟乐圣母堂，米兰

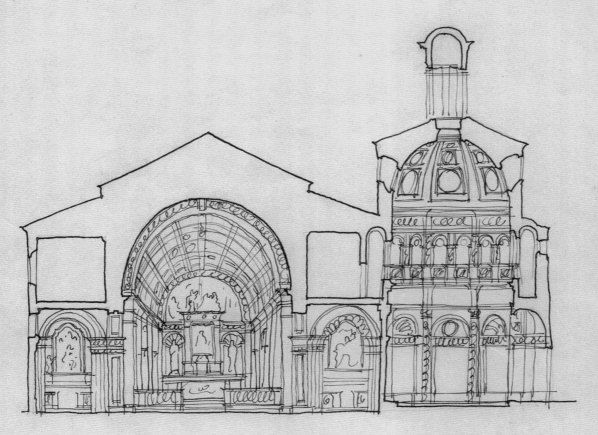

CHANCEL IN PERSPECTIVE 圣坛透视图 SACRISTY 圣器室

SAN SATIRO, MILAN. 圣沙弟乐圣母堂，米兰

(SIMPSON)
(辛普森)

FRONT ELEVATION 前立面图

S.M. DI SAN SATIRO, MILAN
圣沙弟乐圣母堂，米兰
(VALERI)
(瓦莱里)

BRAMANTE'S DESIGN FOR
THE S. SATIRO, FACADE
布拉曼特设计的圣沙弟乐圣母堂正立面图
—(IN LOUVRE MUSEUM)—
—（收藏于卢浮宫）—

THE CAPITAL ROME
卡皮托利尼博物馆

FROM FLETCHER HISTORY OF ARCHITECTURE
摘自《弗莱彻建筑史》

Michelangelo 米开朗琪罗* 　*Michelangelo，米开朗琪罗（1475–1564），文艺复兴时期杰出的雕塑家、建筑师、画家和诗人，与达·芬奇和拉斐尔并称"文艺复兴艺术三杰"。

The Capital, Rome 卡皮托利尼博物馆，罗马

PALAZZO VENDRAMINI 文德拉明宫
VENICE. 威尼斯

The Venetian School 威尼斯学派

Palazzo Vendramin, Venice 文德拉明宫，威尼斯

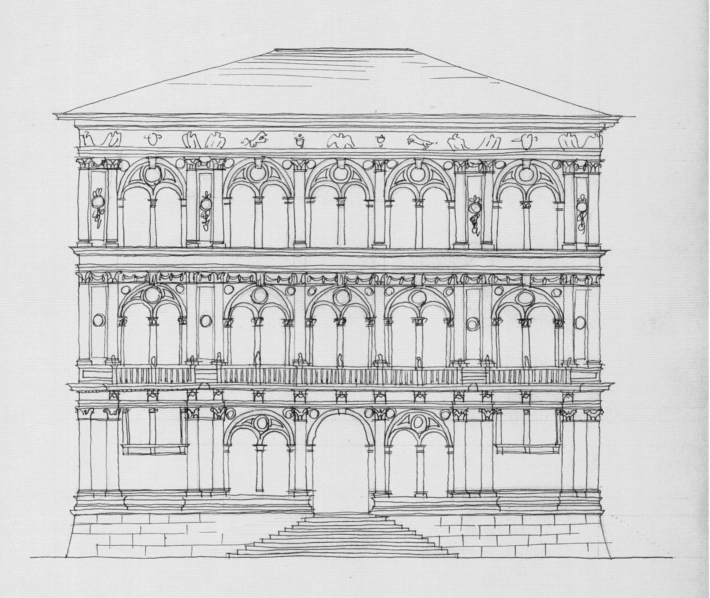

PALAZZO VANDRAMINI CALERGI 文德拉明宫

VEINCE 威尼斯

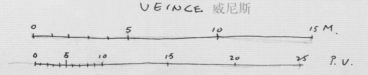

SCUOLO S. MARCO

By Martino Lombardo

1485 - (Gromort & Simpson)

(Muntz)

Adolphe Lauce says " - the structure is in brick masonary while the decoration of marble, in other words the parts of the building which shows are only a covering brought in at the end, the edifice itself has disappeared. One feels that the facade was not much more to Martino than the canvas is to the painter - a surface to be covered according to caprice. "

Let us add that the facade with its foreshortened lions and false arcades, caissoned (which recalls the apse of the S. Satiro, Milan) has something (unexpected) Striking about it.

(Simpson)

Attributed to Martino and Son Moro. Two of the windows on the first floor bear a strong resemblance to the central windows of the Porta de' Borsari, and were, if the date 1485 generally given be correct, the first to have pilasters at the sides and an entablature and pediment above. The semicircular gables over them recalls St. Mark's.

(Photo)

Horizontal accent dominant. Base has Gothic feeling. Ornament on tympanum recalls crokkets. In arabests they are usually renceaus instead of the cantelabra shafts.

Now Municipal hospital.

Scuola Grande di San Marco, Venice 圣马可学校，威尼斯

圣马可学校

马蒂诺·隆巴尔多作品
1485 年（格罗莫尔 & 辛普森）

（芒茨）

阿道夫·兰斯说："建筑由砖石搭建，而大理石的装饰（面），也即建筑展现出来的部分，不过是最后添上的覆盖层，而建筑本身已经消失。建筑物的立面对于马蒂诺来说，与画布对于画家的意义并没有什么区别——用于随心所欲进行创作之物。"

此外，装饰以按透视法缩小的狮子和装饰性假拱廊的立面，加之以凹格镶板的设计（让人联想到米兰圣沙弟乐圣母堂的后堂），具有某种（意想不到的）引人注目之处。

（辛普森）

建筑设计出自马蒂诺和儿子莫罗之手。第二层楼的两扇窗与波萨利门的中央窗有着很强的相似度，且如果一般认为的 1485 年准确的话，是第一批两侧带有壁柱，其上佩以古典柱式顶部和山形墙的窗。

窗上的半圆形山墙让人联想起圣马可大教堂。

（照片）

水平特征显著，基座带有哥特风格，扇形门楣上的饰物让人想起卷叶式凸雕。在阿拉伯式的花纹中，这样的装饰经常取代烛架支柱。

现为市医院。

SCUOLA DI S. MARCO. 圣马可学校

VENICE 威尼斯

—(CICOGNARA)—

—(奇科尼亚拉*)—

* Leopoldo Cicognara,莱奥波尔多·奇科尼亚拉伯爵（1767—1834），意大利考古学家及艺术作家。

Santa Maria dei Miracoli, Venice 奇迹圣母堂，威尼斯

S. M. MIRACOLI

By Pietro Lombardo.

1481 - (Muntz)

1481-1489 - (Gromort)

1480 - (Anderson)

(Muntz)

Facade consists of a great wall pierced with a door, two windows, a huge round window, a three other smaller ones. This predeliction for these round windows which are so awkward is a charcteristic of Venice. All of these are badly bound together. The interior is even more bizzarre and illustrates with what little concern they freed themselves on occassions from the religious traditions, no crucifom plan but a single nave with trancepts, and an altar raised high up, an apse square in plan and a barrel vault in wood. The sculptural decoraion conceals these faults.

(Simpson)

Semi-Renaissance. Won in competition by P.L. in 1481, owes its beauty mainly to its marble vaneers and pattrae, and consequently more to Byzantine and Lombardic traditions thanto the influence of the new movements. The dome which covers the chancel is raised on a drum above pendentives, and extrernally recalls the dome of the St. Mark's. The round top of the nave roof and the semi-circular gable at the west ed appears strange to western eyes, accustomed to pointed gables and pitched roofs, but the latter would have appeared equally strange to the Venetians.

奇迹圣母堂

彼得罗·隆巴尔多作品

1481 年（芒茨）

1481 年 – 1489 年（格罗莫尔）

1480 年（安德森）

（芒茨）

立面包括一堵由门穿透的高墙、两扇窗、一扇巨大的圆形窗以及另外三扇较小的窗。对这些别扭的圆形窗的偏爱是威尼斯的特色之一，所有的这些都不甚协调地连在一起。内部则更为古怪，体现出偶尔摆脱宗教传统时是多么地随意，却只有一个附以耳殿的中殿、一个高祭坛、一个规划为方形的后堂和一个木制的桶形拱顶。雕塑装饰掩盖了这些缺点。

（辛普森）

具有文艺复兴早期特点。彼得罗·隆巴尔多于 1481 年中标，其美感主要归因于大理石薄片镶饰和对称拼砌的大理石花纹，因此受拜占庭和伦巴第传统的影响要大于当时新运动的影响。高坛上方的圆穹顶和穹顶支承拱间加入了鼓形座，从外部上看屋顶的设计让人联想到圣马可大教堂的穹顶。教堂中殿的圆形顶部和西部尽头半圆形的山墙对于习惯了尖山墙和坡屋顶的西方人是陌生的，但是后者对于威尼斯人来说则同样陌生。

(Drawings)

  Plan breaks tradition, without aisles orside chapels. Chancel raised consideragly, good for ceremony. This high end is also Lombard. (S. Zeno at Verona, has one.) Covered by barrel vault ad dome. The <u>pendentive arch is expressed on exterior</u>. The facade accentuates horizontality. Bays of arches are not the same width for adjointing arches. Round roof had been used in Gothic time, following wooden contour inside.

（手绘图）

  设计图突破了传统，小教堂之外不设侧廊。高坛高度上升明显，利于举行仪式。这种把尽头抬高的手法也是伦巴第的风格（维罗纳的圣杰诺也有一个）。（尽头的）上方为桶形拱顶和圆屋顶。穹隅的拱门居于建筑外部。建筑正面强调水平感。拱门的隔间与毗连的拱门宽度不同。圆屋顶早在哥特时期已有使用，屋顶内部为木制结构。

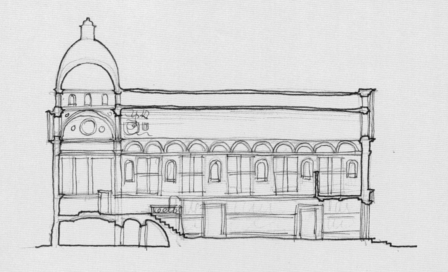

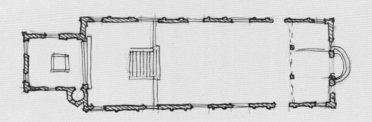

S. M. DE MIRACOLI 奇迹圣母堂

VENICE 威尼斯

PAL. CORÉR-SPINELLI 科纳·斯皮内利府邸

VENICE 威尼斯

—(ITALIA ARTISTICA)—
—（意大利艺术）—

Palazzo Corner Spinelli, Venice 科纳·斯皮内利府邸，威尼斯

S. ZACCARIA. 圣匝加利亚教堂
VENICE 威尼斯

—(CICOGNARA)—
—(奇科尼亚拉)—

San Zaccaria, Venice 圣匝加利亚教堂，威尼斯

S. ZACCARIA VENICE
圣匝加利亚教堂　威尼斯
—(CICOGNANA)—
—(奇科尼亚拉)—

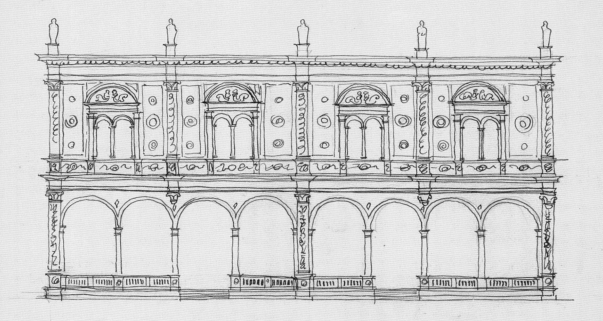

PALAZZO DEL CONSIGLIO 议会大厦
VERONA · 维罗纳

—（REYMOND）—
—（雷蒙）—

Palazzo del Consiglio, Verona 议会大厦，维罗纳

PALAZZO BEVILACQUA, BOLOGNA

(Muntz)

Begun 1481.

With fascetted rustication on facade, more curious than beautiful, court however very distinguished. Two stories of arcades arranged so that two arches of the upper story corresponed to one of the lower. The frieze is ornamented with half length fibures terminating in renceau. A superb cornice. Precise details are lacking of the work of Francesco Rrancia.

This work was credited to Gaspare Nardi. But recent research found that he was only a mason.

(Simpson)

Window on second floor is like Florentine but has not florentine feeling. In the court the bays on second floor is doubled th at of the first.

(Gumaer)

One very bad thing here in Bologna is the use of bad stone for fine detail. The fascetts are not very prominent and are varried. Delicate proportion and detail in windows. Splayed base and projecting seat. At end is Venitian pilaster.

Palazzo Bevilacqua, Bologna 饮泉宫．博洛尼亚

饮泉宫，博洛尼亚

（芒茨）

1481 年开始建造。

正立面为许多刻成小面的粗面石，说不上很美，显得有些怪异，然而中庭卓然不群。此建筑设有两层拱廊，上层的两个拱门与下层的一个拱门恰好对应。雕带饰以半长叶漩涡饰。飞檐的设计独具匠心。细节的设计缺少弗朗切斯科·弗朗西亚*的设计风格。

人们一直认为这栋建筑出自于加斯帕雷·纳尔迪之手，然而最近的研究显示该人只是个石匠。

* Francesco Francia，弗朗切斯科·弗朗西亚（约1450—1517），意大利画家。

（辛普森）

二层窗户与佛罗伦萨样式相似，但并无佛罗伦萨的味道。中庭二层隔间是一层的两倍。

（古米尔）

用粗石块展现细节的精美是博洛尼亚建筑的一大缺陷。组成建筑的粗面石形态并不引人注目，且形态各异。窗户的设计注重比例和细节的雕饰。基部为斜面，底座向前伸出，拐角处配有威尼斯式壁柱。

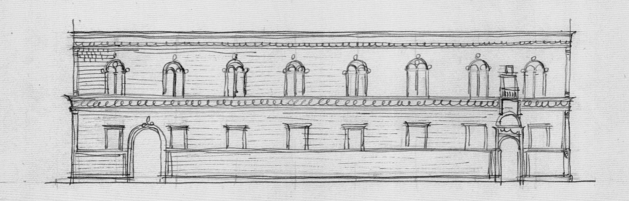

PALAZZO BEVILACQUA 饮泉宫

BOLOGNA. 博洛尼亚

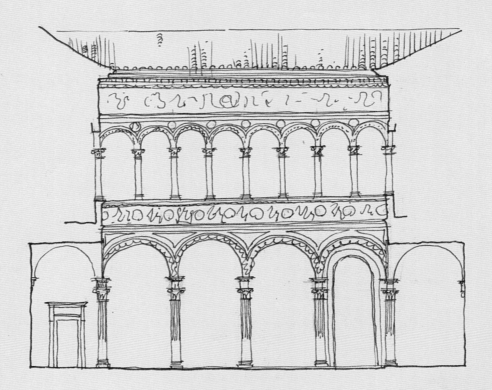

PALAZZO BEUILACQUA   BOLOGNA
COURT
饮泉宫 博洛尼亚 中庭
—(PHOTO)—
—(照片)—

Palazzo Fava Ghisilieri, Bologna 法瓦大殿，博洛尼亚

PALAZZO FAVA. 法瓦大殿
BOLOGNA. 博洛尼亚

LOMBARD SCHOOL

(Mintz)

<u>Characteristics</u> of style in Lombardy during first third of XV century are:
    (1) <u>Persistance of Gothic</u>,
    (2) <u>Picturesqueness</u>,
    (3) <u>Profusion of ornament</u>,
    (4) <u>Use of terra cotta</u> (repitition of ornament)
    (5) <u>Retention of pointed arch</u>

The Florentine set the movement going in the North, but once started the Lombards made it their own and produced works which were inspired by the Romanesque buildings, Santa Eufemia at Pavia, the Abbey of Chiaravalle the Church of S. Gottardo at Milan. The return to older forms such as Romanesque or Early Christian frequently preceeded the study of the Classic monuments and aided in the transition. The Roman monuments in the North rather incorrect as compared with those in central Italy.

The importance of the construction of the Cathedral at Milan frequently overshadows other monuments such as the Castle of the Sforza begun in 1450. Ospedale Maggiore, le Laqaretto, the Portinari Chapel and a series of palaces. Brunelleschi was called to Milan by Philippo Maria. Outside of him the principal Florentine apostle was Filarette, sculptor and architect.

Filarete not stong enough alone to have accomplished the Renaissance in the North movement greatly helped by Michelozzo, who was given charge of work on Palace which Francesco Sforza gave to Cosmo Medici in 1456.

伦巴第学派

(芒茨)

十五世纪的前三十年中伦巴第的风格特点表现为：

(1) 保持了哥特式的风格；

(2) 精美如画；

(3) 富于装饰；

(4) 使用赤土陶（重复的装饰）；

(5) 保留尖拱顶。

佛罗伦萨人首先在北部发起了运动，但伦巴第人后来居上，发展出自己的风格，受罗马式建筑启发设计建造了一系列建筑，例如帕维亚的圣埃乌费米亚教堂、齐亚拉瓦莱修道院、米兰的圣哥达教堂。在对古典建筑进行细致研究前往往有向罗马式或早期基督教等早期建筑形式的回归，推动了风格的过渡。相比于那些意大利中部的古迹，北方的罗马建筑（在这一点上）体现得不是很好。

由于米兰大教堂具有重要意义，它的建造经常使其他的建筑名胜失去光彩，例如 1450 年开始建造的斯福扎城堡、米兰市立医院、传染病院、波尔蒂纳里礼拜堂和一系列的宫殿。布鲁内莱斯基被菲利普·马利亚·维斯康提*征召到米兰。除他之外，最主要的佛罗伦萨画派的代表人物是雕塑家、建筑家菲拉雷特。

菲拉雷特本人的力量还不足以实现北方的文艺复兴运动，他得到了米开罗佐*的巨大帮助。米开罗佐后来负责了弗朗切斯科·斯福扎于 1456 年交给科西莫·德·美第奇*的宫殿建筑工作。

---

* Filippo Maria Visconti，菲利普·马利亚·维斯康提（1392—1447），15世纪米兰公爵，米兰大主教。

* Michelozzo，米开罗佐（1396—1472），意大利建筑师，雕塑家。

* Cosimo de' Medici，科西莫·德·美第奇（1389—1464），文艺复兴时期著名的佛罗伦萨僭主，大商人。

FILARETE

Antonia di Pietro Averrulino, or Averlino, called Filarete. (Philareti from Greek "friend of virtue).

<u>Dates</u>   Born 1400, Florence.
　　　　　Died 1469 (Muntz)
　　　　　　　1465 (Simpson).

(Muntz)

Woked on Baptistry dors with Ghiberti. Mediocre as sculptor but obtained favour of Egenius IV during sojourn of Pope in Florence and obtained commission for bronze door of St Peter's.

Accused at geginning of Pontificate of Nicholas V, 1448-49 of having stolen the head of John the Baptist. He was imprisoned and finally banished from Rome. Came to Milan where he was in 1451 put in relation with the Sforza by the Medici. He worked on the Cathedral and on Sforza castle and finally the Hospital.

Begun construction of the cathedral of Bergamo which was finished long after his death by carlo Fontana.

At Milan in service of the duke, Francesco Sforza, that he composed his treatise on architecture. After the death of Francesco, Filarete seems to have returned to Florence, then to Rome where Vasari makes him die at the age of 69.

(Simpson)

Born 1400, Florence, remained there till 1433. Next 12 years in Rome, where, amongst other works, he executed the fine bronze doors for St. Peters 1433-1443, which now form the central entrance to the church. From Rome he went to Venice, and there designed and carried out a procession silver

菲拉雷特

菲拉雷特,又名安东尼奥·迪·彼得罗·阿韦利诺、阿韦利诺。
(Philareti在希腊语中意为"品德高尚的朋友"。)

日期　生于 1400 年,佛罗伦萨。

　　　　卒于 1469 年(芒茨)

　　　　　　 1465 年(辛普森)。

(芒茨)

和洛伦佐·吉贝尔蒂*共同完成了洗礼堂的门的建筑。作为一个平庸的雕塑家,却在罗马教皇犹金四世逗留佛罗伦萨期间得到了其青睐,并且获得了建造圣彼得教堂铜门的工作。

1448 – 1449 年,在教皇尼古拉五世刚登基时,他被指控偷盗施洗者约翰的头颅而入狱,并最终被流放到罗马以外。后到达米兰并于 1451 年由美第奇家族引荐给斯福扎。他先后参与了大教堂、斯福扎城堡以及医院的建造。

开始建造贝加莫大教堂,这座教堂在他死后很长时间才由卡洛·丰塔纳*完成。

在米兰服务于公爵弗朗切斯科·斯福扎期间,菲拉雷特完成了建筑专著。在弗朗切斯科死后,菲拉雷特似乎返回了佛罗伦萨,然后去了罗马。据瓦萨里记载享年 69 岁。

* Lorenzo Ghiberti,洛伦佐·吉贝尔蒂(1378—1455),意大利文艺复兴初期雕塑家。

* Carlo Fontana,卡洛·丰塔纳(1634或1638—1714),意大利建筑师。

(辛普森)

1400 年生于佛罗伦萨,并一直生活至 1435 年。接下来的 12 年居住于罗马,1433 – 1445 年修建了圣彼得大教堂精美的铜门,现为教堂的中心入口。

cross for the Cathedral of Bassano.   Then finished with a letter of intruduction from Piero de Medici to Francesco Sforza, he started for Milan, where he died 1465 or 1469.

后从罗马来到威尼斯，设计并建造了巴萨诺大教堂的一列银色十字架。得到皮耶罗·德·美第奇[*]写给弗朗切斯科·斯福扎的推荐信后启程来到米兰，并于 *1465* 年（或 *1469* 年）在米兰去世。

* Piero di Cosimo de'Medici，皮耶罗·迪·科西莫·德·美第奇（1416—1469），科西莫·德·美第奇之子，1464—1469年间统治佛罗伦萨。

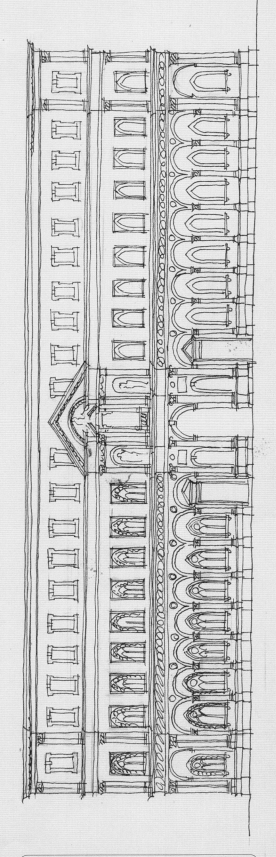
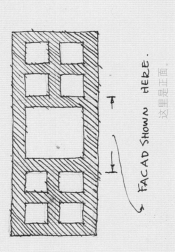

The Ospedale Maggiore, Milan 米兰市立医院

PART II
第二部分

Architecture Notes:
A Geographical Perspective
建筑笔记：地域的视角

ROME 罗马
Palazzo Venezia & Church of S. Mark
威尼斯宫和圣马可教堂

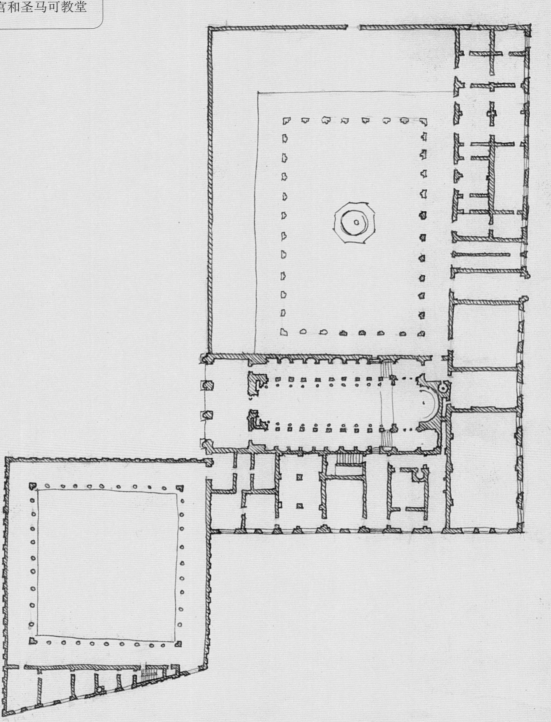

PALAZZO VENEZIA & CHURCH OF S. MARCO 威尼斯宫和圣马可教堂
ROME 罗马
—(LETAROUILLY)—（勒塔鲁伊）—

DETAIL 细部

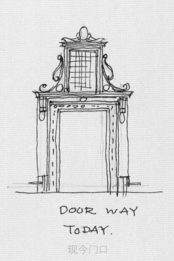

DOOR WAY TODAY.
现今门口

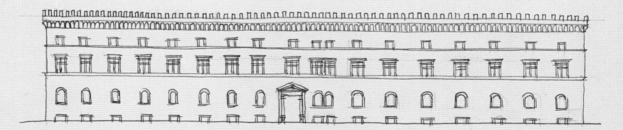

FAÇADE 正立面

PALAZZO VENZIA 威尼斯宫
GRAND PALACE. 大宫

PALAZZO VENEZIA ROME 威尼斯宫，罗马
THE SMALL PALACE 小宫
—(LETAROUILLY)—
—（勒塔鲁伊）—

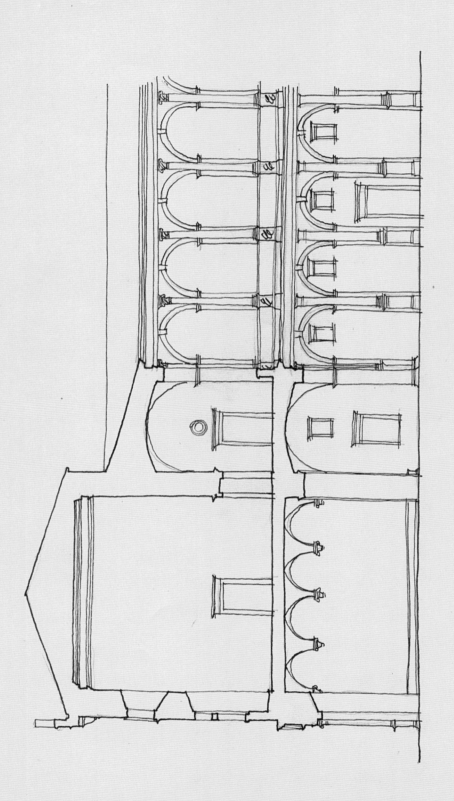

Palazzo Venezia Rome
威尼斯宫，罗马
Section Grand Palace
大宫剖面图

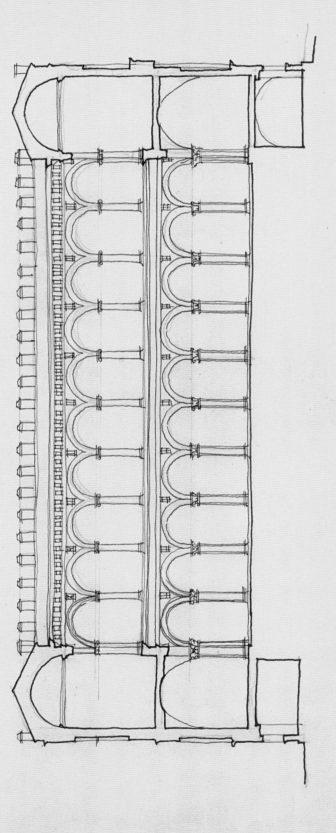

PALAZZO VENEZIA ROME 威尼斯宫，罗马
SECTION PETITE PALACE. 小宫切面图

S.M. DELL' ANIMA. 圣马利亚灵魂之母教堂

ROME. 罗马

—(LETAROUILLY)—
—（勒塔鲁伊）—

Santa Maria dell'Anima
圣马利亚灵魂之母教堂

S. M. DEL ANIMA   ROME

(Letarouily)

This church with adjoining hospital begun in 1400 from funds left by Giovanni di Pietro, a native of Flanders.   Added to by Germans who still maintain it.

Church takes its name from an ancient figure found on the site representing the Virgin between two faithful souls.   A copy of this is placed in the pediment over the main entrance.

The facade believed to have been constructed in 1522.   Letarouily thinks it earlier, perhaps 1500.   It is attributed to Antonio da San Gallo, the elder.   Some critic attribut oto him only the entrance doors. In a note it is suggested that the upper part of the facade is by Perruzzi who is architect of Tomb of Adrian VI in interior.   Vasari says Bramante was consulted in regard to church.

(Drawing)

No trancept.   Only 4 bays.   Side aisles with semi-elliptical chapels. The choir has width of Nave, ends with apse.

Facade has three story, 3 bays.   First story one large central doorway, framed with Corinthian order crowned with triangular pediment.   Two side doors smaller, but with   Pediment.   2nd story has one arch in each bay, 3rd has round window in centre.   No heavy crowning unit.   The 3 bays are divided by Corinthianesque pilasters in all three storeys.

(Gumaer) Nothing in facade suggests section.   The spalayed wall is taken care of by varying the depths of the chapels, thus getting a rectangular plan.

It is possible that only the 3 doorways are by San Gallo.   The facade as a whole is rather Mechanical.

圣马利亚灵魂之母教堂，罗马

（勒塔鲁伊）

教堂及其毗邻的医院始建于 1400 年，由佛兰德斯人乔瓦尼·彼得资助。后由德国人继续资助，现仍由德国人进行维护。

教堂由遗址上发现的一尊古老的塑像得名，塑像形象为两位虔诚者簇拥居中的圣母。仿制品安放在正门顶的三角楣饰上。

正立面据称建于 1522 年。而勒塔鲁伊则认为可能更早，或许在 1500 年。建筑由老安东尼奥·达·圣加罗负责建造设计。一些批评家认为只有大门是他设计的。根据一份笔记记载，建筑正面的上半部分由负责阿德里安六世）墓室内部建筑的佩鲁齐完成。瓦萨里则认为布拉曼特也参与了教堂的建筑。

（手绘图）

没有耳堂似乎更常见，这样说也不错，只有四个隔间。侧廊有半椭圆形的小礼拜堂。唱诗班坐席和教堂中殿宽度相同，末端为半圆形后殿。

正立面有三层，三个隔间。第一层有一个很大的中央门廊，饰以科林斯柱，顶端为三角形顶饰。两边的门较小，但是也有三角形顶饰。二层每个隔间均有一个拱门。三层中间为圆形窗户。顶端没有沉重的装饰元素。所有三层的三个隔间都由科林斯壁柱隔开。

（古米尔） 立面完全没有反映剖面形式。尽管两边的墙体各自向外倾斜（类似梯形的两腰），但由于拓宽了一部分礼拜堂的深度，教堂（的内部）仍呈长方形。

有可能仅有三个门廊是由圣加罗设计建造，正面整体的感觉较为呆板。

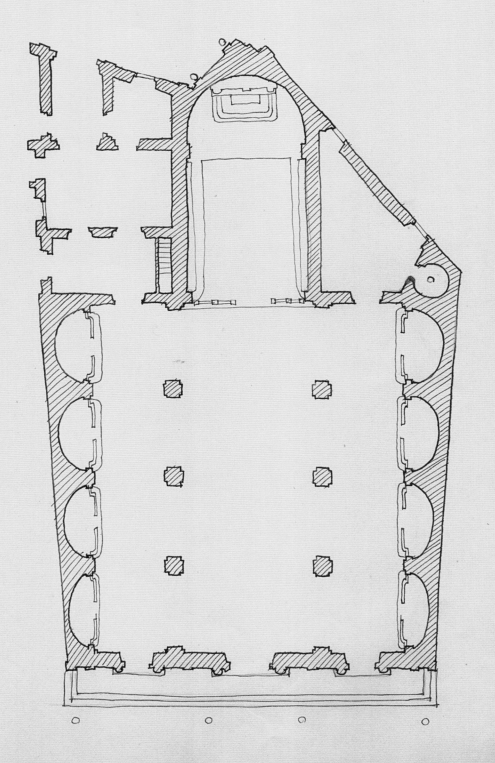

S. M. DELL' ANIMA. 圣马利亚灵魂之母教堂

—(LETRAOUILLY) — —（勒塔鲁伊）—

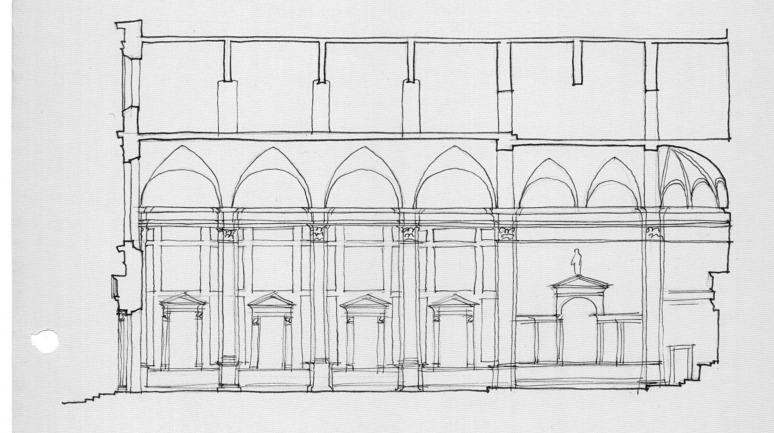

S.M. DELL' ANIMA.
圣马利亚灵魂之母教堂

—(LETAROUILLY)—
—(勒塔鲁伊)—

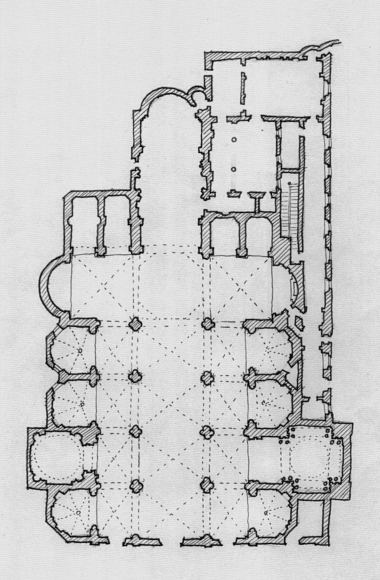

S.M. DEL POPOLO 人民圣母教堂

ROME 罗马

—(LETAROUILLY)—
—（勒塔鲁伊）—

Santa Maria del Popolo
人民圣母教堂

S.M. DEL POPOLO 人民圣母教堂

ROME 罗马

—(LETAROUILLY)—
—（勒塔鲁伊）—

S. PIETRO IN MONTORIO 蒙托利尔的圣彼得教堂

ROME 罗马

—(LETAROUILLY)—
—（勒塔鲁伊）—

San Pietro in Montorio
蒙托利尔的圣彼得教堂

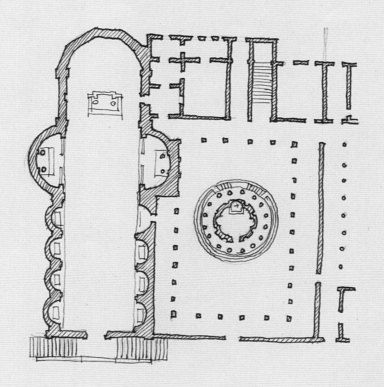

S. PIETRO IN MONTORIO 蒙托利尔的圣彼得教堂

ROME 罗马

—(LETAROUILLY)—

—（勒塔鲁伊）—

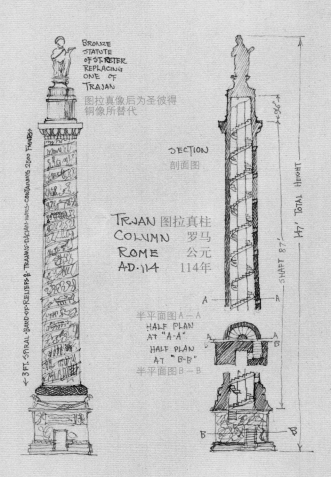

Trajan's Column 图拉真柱

MILAN 米兰
The Door of the Medici Bank
美第奇银行的门

The Door 美第奇银行的门
of the Medicci Bank

Milan 米兰

—(Artistica Italiana)—
—（意大利艺术）—

Basilica of San Lorenzo
圣洛伦佐教堂

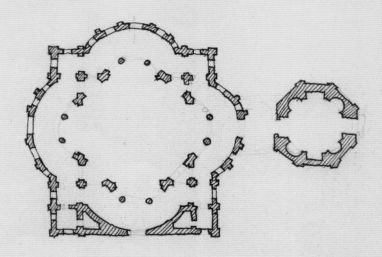

S. LORENZO, MILAN. 圣洛伦佐教堂，米兰

FROM WHICH BRAMANTE GOT SUGGESTION FOR HIS SCHEME FOR THE ST. PETER'S.

在设计圣彼得大教堂时，布拉曼特从中得到启发。

CHURCH AT ABBIATEGRASSO
ENTRANCE PORCH
阿比亚泰格拉索教堂,入口处的门廊
— (SIMPSON) —
—(辛普森)—

Abbiategrasso
阿比亚泰格拉索教堂

Chapel of San Pietro Martire
殉道者圣彼得小教堂

FLORENCE 佛罗伦萨
Villa Careggi 卡勒吉别墅

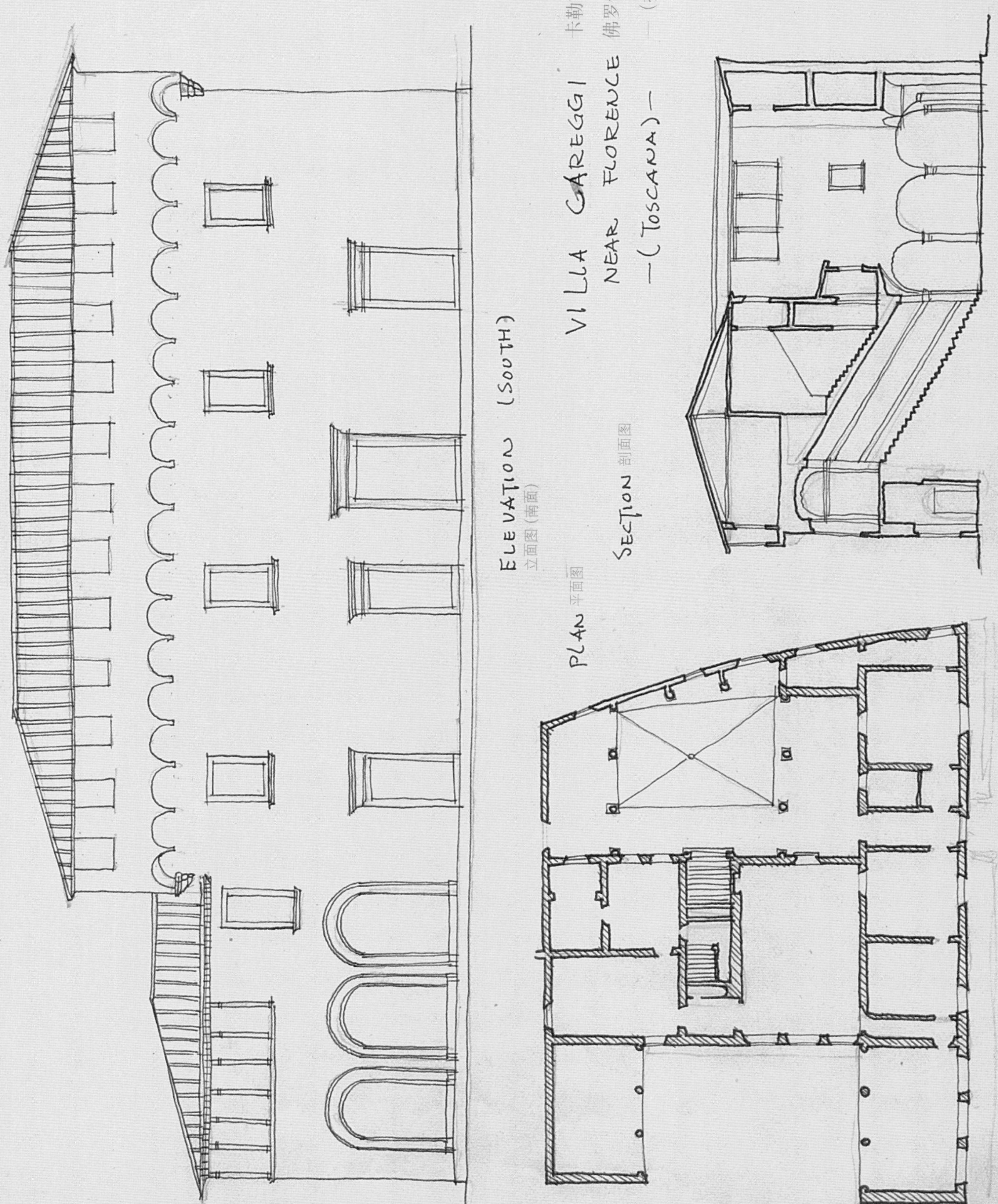

VILLA CAREGGI 卡勒吉别墅
NEAR FLORENCE 佛罗伦萨附近
—(TOSCANA)— —(托斯卡纳)—

ELEVATION (SOUTH) 立面图(南面)

SECTION 剖面图

PLAN 平面图

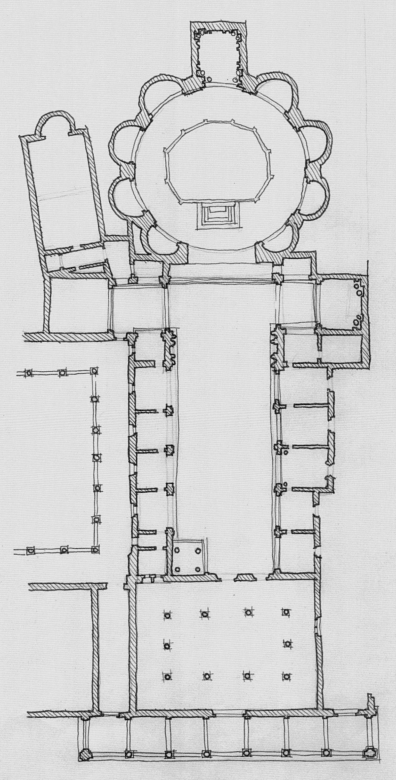

Santissima Annunziata 圣母领报大殿

Palazzo Medici Riccardi 美第奇-里卡迪宫

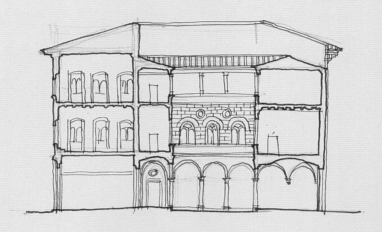

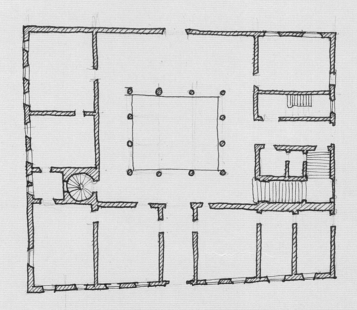

美第奇-里卡迪宫，佛罗伦萨
PALAZZO RICARDI. FLORENCE
(TOSCANA) (托斯卡纳)

PALAZZO RICCARDI, FLORENCE.
From
美第奇－里卡迪宫，佛罗伦萨
Fletcher - HISTORY OF ARCHITECTURE.
摘自《弗莱彻建筑史》

PALAZZO RICCARDI
美第奇-里卡迪宫
(GROMORT)
——（格罗莫尔）——

Basilica of Santa Croce
圣十字教堂回廊

DOORWAY CLOISTER OF
ST CROCE 门廊，圣十字教堂回廊
FLORENCE 佛罗伦萨
(GRANDJEAN) (格朗让)

The Marsupini Tomb 马苏匹尼墓
in S. Croce, Florence 圣十字教堂，佛罗伦萨

The Marsupini Tomb in S. Croce
圣十字教堂马苏匹尼基

Chapel of Crucifix (San Miniato) 圣体小堂（圣米尼亚托教堂）

FRONT ELEVATION 前立面   REAR ELEVATION 后立面

PLAN 平面

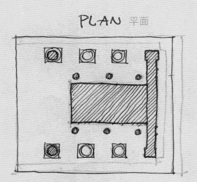

CHAPEL OF CRUCIFIX (S. MINIATO) FLORENCE
圣体小堂（圣米尼亚托教堂），佛罗伦萨
(TOSCANA)
（托斯卡纳）

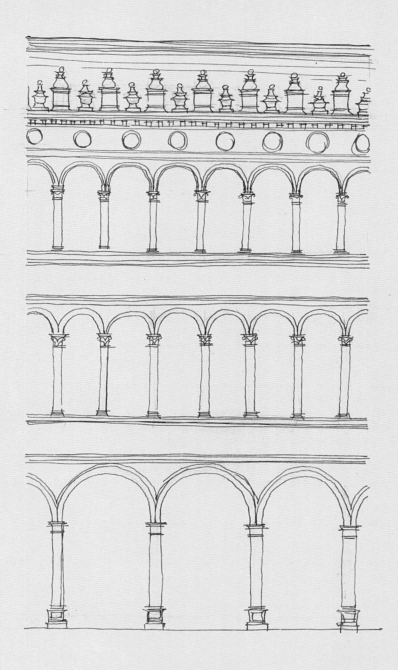

PROCURATIA VECCHIE 旧行政长官官邸大楼

PARTIAL ELEVATION 局部立面图

VENICE 威尼斯
Procuratie Vecchie 旧行政长官官邸大楼

S. GIO. EVANGELISTA.
圣约翰福音教堂
VENICE.
威尼斯

Chiesa di San Giovanni Evangelista
圣约翰福音教堂

SQUOLA S. ROCCO.
圣洛可大会堂

Scuola Grande di San Rocco
圣洛可大会堂

SQUOLA S. ROCCO
圣洛可大会堂
VENICE.
威尼斯

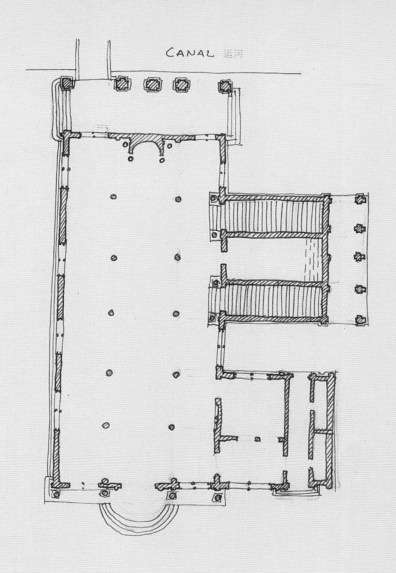

SQUOLA S. ROCCO 圣洛可大会堂
VENICE. 威尼斯

PALAZZO DAIRO 达里欧宫

VENICE 威尼斯

—（PHOTO）—
—（照片）—

Palazzo Dairo 达里欧宫

CHURCH OF S. SPIRITO
圣灵教堂

BOLOGNA.
博洛尼亚

BOLOGNA 博洛尼亚
Church of San Spirito 圣灵教堂

FRONT ELEVATION
前立面

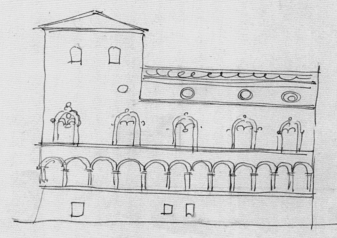

SIDE ELEVATION
侧立面

PALAZZO CARRACCI BOLOGNA
卡拉奇宫,博洛尼亚

Palazzo Carracci 卡拉奇宫

PALAZZO PALLAVICINI. 帕拉维奇尼宫

BOLOGNA、博洛尼亚

Palazzo Pallavicini 帕拉维奇尼宫

PALAZO DELL'ARTE DEGLI
STRAZZAROLI
斯特拉察罗里宫
BOLOGNA.
博洛尼亚

Palazzo degli Strazzaroli
斯特拉察罗里宫

COPUS DOMINI BOLOGNA
圣体教堂，博洛尼亚
DOORWAY.
门廊

Corpus Domini
圣体教堂

BALL FLOWER — (DECORATED)
球心花饰（盛饰式的）

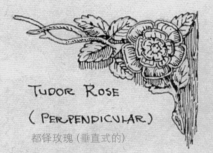

TUDOR ROSE
(PERPENDICULAR)
都铎玫瑰（垂直式的）

CROCKETS 花形浮雕

EARLY ENGLISH
早期英国式的

DECORATED
盛饰式的

PERPENDICULAR
垂直式的

ENGLISH GOTHIC ORNAMENTS.
英国哥特式装饰

> **VERONA 维罗纳**
> English Gothic Ornaments
> 英国哥特式装饰

PALAZZO COMMULE 市政府

BRESCIA 布雷西亚

BRESCIA 布雷西亚
Palazzo Communale 市政府

S.M. DEI MIRACOLI 奇迹圣母堂

BRESCIA. 布雷西亚

Santa Marie dei Miracoli
奇迹圣母堂

MONTE DI PIETA, 典当行

BRESCIA . 布雷西亚

Monte di Pietà
典当行

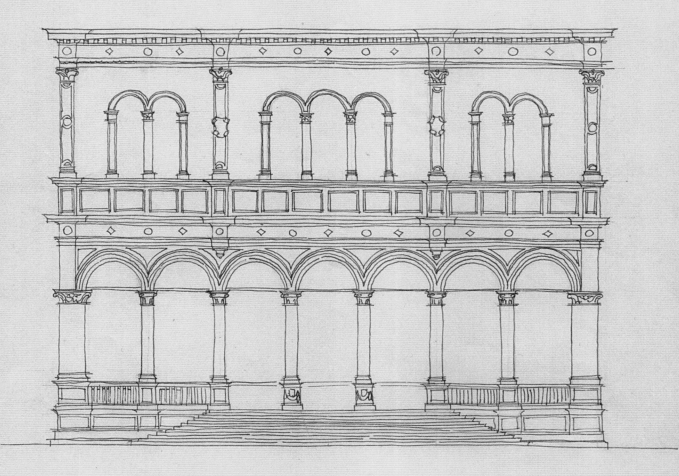

LOGGIA DEL CONSIGLIO
市政会凉廊
PADUA
帕多瓦

—(HAUPT)—
—(豪普特)—

PADUA 帕多瓦
Loggia del Consiglio 市政会凉廊

FERRARA 费拉拉
Palazzo Diamanti 钻石宫

PALAZZO SCHIFANOIA
FERRARA
席法诺亚宫，费拉拉
DOOR WAY.
门廊
—（BAUM）—
—（鲍姆）—

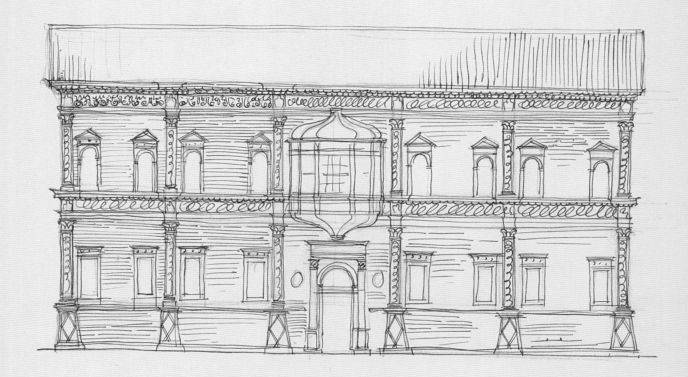

PALAZZO ROVERELLA 罗维戈宫

FERRARA 费拉拉

—(PALAST-ARCHITEKTUR)—
—(皇宫建筑)—

Palazzo Roverella 罗维戈宫

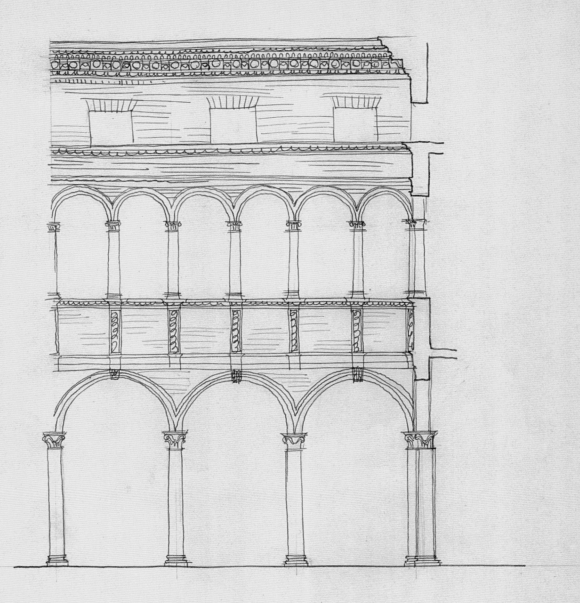

Palazzo Costabili 科斯塔比利宫

PALAZZO PODESTA 行政长官官邸

PERUGIA 佩鲁贾

—PHOTO—
—照片—

PERUGIA 佩鲁贾
Palazzo Podesta 行政长官官邸

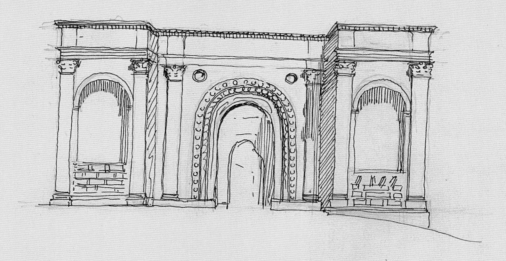

PORTA URBIRA DI S. PIETRO 圣彼得门

PERUGIA 佩鲁贾

DUCCIO 杜乔 1473

(BAUKUNST)
（建筑）

PALAZZO DEL PRETORIO 法庭宫
PIENZA 皮恩札

(TOSCANA)（托斯卡纳）

PIENZA 皮恩札
Palazzo del Pretorio 法庭宫

IL Duomo di Pienza 皮恩札主教堂

IL DUOMO
PIENZA
皮恩札主教堂
(TOSCANA)
(托斯卡纳)

Palazzo Picolomini 皮克罗米尼宫

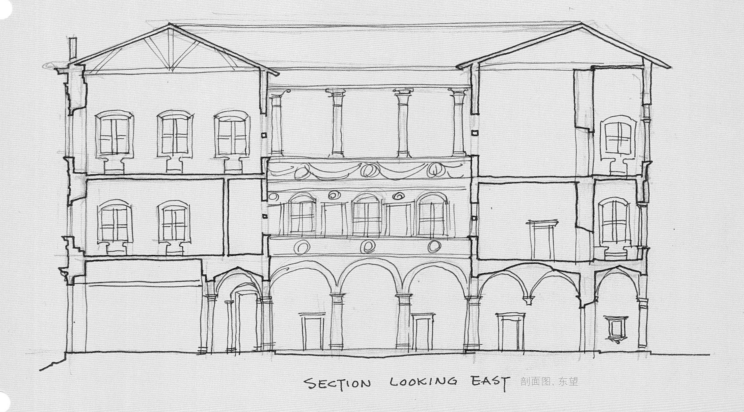

SECTION LOOKING EAST 剖面图、东望

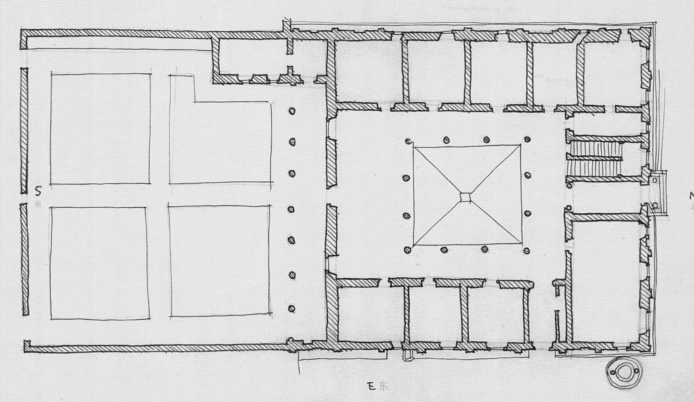

PLAN 平面图

PALAZZO PICOLOMINI PIENZA 皮克罗米尼宫 皮恩札
(PALAST-ARCHITECTURE TOSCANA) (皇宫建筑 托斯卡纳)

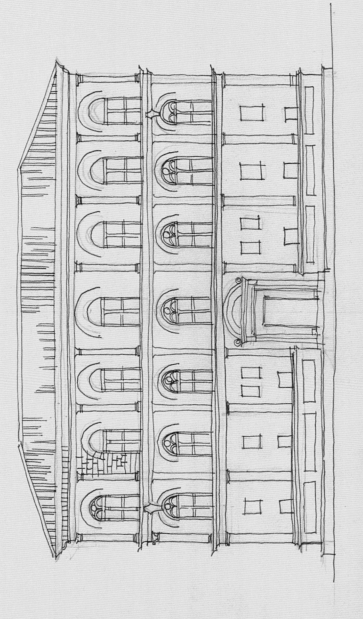

PALAZZO PICOLOMINI 皮克罗米尼宫
PIENZA 皮恩札

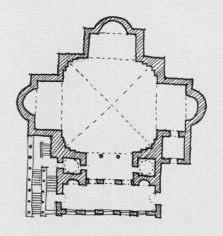

PLAN 平面图

SAN SEBASTIANO, MANTUA.
圣塞巴斯蒂亚诺，曼图亚

MANTUA 曼图亚
Chiesa di San Sebastiano
圣塞巴斯蒂亚诺

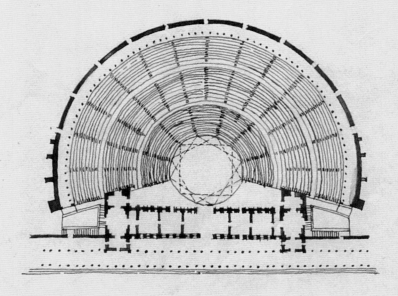

GREEK THEATRE
AFTER COCKERELL
摹科克雷尔*绘希腊剧场

*Charles Robert Cockerell, 科克雷尔（1788—1863），英国建筑师、考古学家、作家。

SHOWING SETTING OUT AND CONSTRUCTION OF A TYPICAL THEATRE UPON THE AUTHORITY OF VITRUVIOUS, POLLUX, AND OTHERS. — THE ORCHESTRA OR DANCING PLACE WAS OCCUPIED BY THE CHORUS AND DANCERS ONLY. THE INSIDE BOUNDARY OF THE SEATING FORMED ABOUT TWO-THIRD OF A CIRCLE. THE STAGE OR SKÉNÉ FORMED DRESSING ROOMS FOR THE ACTORS, AND WAS RAISED CONSIDERABLY ABOVE THE ORCHESTRA.

展现了维特鲁威、波勒克斯及其他权威人士所设计的剧场的典型布局和构造。合唱队席（即表演区）为合唱队和舞蹈演员专用。观众席的范围占据2/3个圆形。舞台上有演员更衣室，高过表演席。

> **OTHERS 其他**
> **Greek Theatre 希腊剧场**

CHÂTEAU DE BLOIS
布卢瓦城堡

—FLETCHER's—History of Architecture.
—《弗莱彻建筑史》

Chateau de Blois, Paris
布卢瓦城堡，巴黎

圣厄斯塔什教堂，巴黎

FROM FLECTCHER
《弗莱彻建筑史》

Saint-Eustache, Paris
圣厄斯塔什教堂，巴黎

CHATEAU DE CHAMBORD
香波尔城堡

From "FLETCHER'S"
《弗莱彻建筑史》

Chateau de Chambord, Paris
香波尔城堡，巴黎

DORMER-WINDOW
CHATEAU DE CHAMBORD
天窗，香波尔城堡

The Colonnade of the Louvre, Paris
卢浮宫柱廊，巴黎

ELEVATION 立面图

ENTRANCE TO COUR HENRI IV.
FONTAINEBLEAU,
枫丹白露宫

FROM — "Palais de FONTAINEBLEAU" —
来自—"枫丹白露宫"—

Chateau de Fontainebleau, Paris
枫丹白露宫，巴黎

THE PANTHEON, PARIS.
先贤祠，巴黎

The Pantheon, Paris
先贤祠，巴黎

CHORAGIC MONUMENT OF LYSICRATES
ATHENS, GREECE.
奖杯亭
雅典，希腊

Choragic Monument of Lysicrates, Athens
奖杯亭，雅典

Temple at Philae
Entrance Court Showing Pylons.
菲莱神庙，宫殿入口的塔门

Temple at Philae
菲莱神庙

Santa Sophia　　　圣索菲亚大教堂
Constantinople　　君士坦丁堡
From the University Prints　源自校园印刷品

Illustrations of the Composition of Mass.

Please find examples in some book other than Univ. Prints.

San Sophia, Constantinople
圣索菲亚大教堂，君士坦丁堡

Palazzo Picolomini, Siena 皮克罗米尼宫，锡耶纳

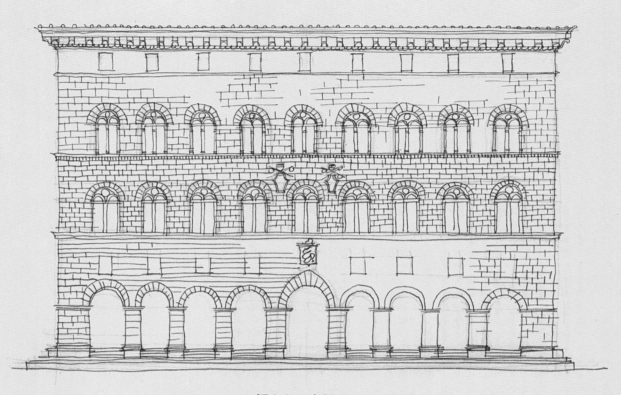

ELEVATION 立面图

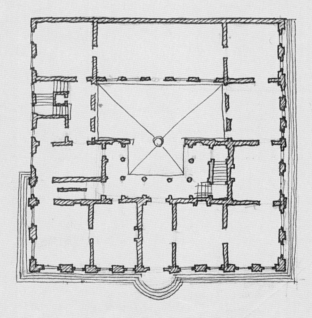

PLAN 平面图

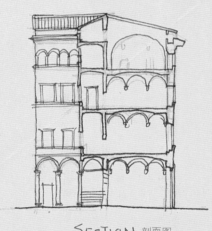

SECTION 剖面图

PALAZZ PICOLOMINI, SIENA 皮克罗米尼宫，锡耶纳
(GRANDJEAN) (格朗让)

APPENDIX

附录

Manuscript of Liang's In-class Notes on History of Architecture

梁思成建筑史课堂笔记原稿

1924年，梁思成和林徽因共赴美国费城宾夕法尼亚大学留学，梁思成入读建筑学院。开学后，梁思成旁听了建筑史教授艾尔弗雷德·H. 古米尔 (Alfred H. Gumaer) 为二年级学生开的一门课 (课程号Architecture 44, 课程名称History of Architecture)。或许是受到父亲的影响，他越听越感兴趣，于是找到教授请求允许他提前一年选修此课。古米尔教授非常喜欢这个热爱建筑史的学生，同意了他的请求。附录文字为梁思成先生修读这门课程所整理的课堂笔记，因"文革"期间笔记屡遭厄运，个别地方有缺页。

Architecture 44.

HISTORY OF ARCHITECTURE
Lectures by Professor Gumaer

Lecture 1, September 28, 1925.

RENAISSANCE ARCHITECTURE

<u>What is
Renaissance</u> The word Renaissance is originally a
French word - meaning "rebirth".
This term is used to designate the revival of classic
culture at the beginning of the fifteen century.
The real foundation of the movement was individualistic.
Simmon called it the "self conscious freedom". With
Renaissance came the idea of this "self Conscious freedom" - ability for the individual to think for himself,
- this involved the revolution again st the Church.
"Renascence" is the English meaning the same thing.
People refering to this period, frequently, critics especially, refered to it as a period of imitation.
It is true to a certain extend, but they seemed to loose
sight of how much the Romans depended on the Greeks,

-2-

even more similarly than the Renaissance to the Classics. So many things change and so many remain that it is impossible to illimate the periods of architecture.

Completeness is the most Phenominal characteristic of this period; it extended from Florence, France,..... to England.

<u>How Renaissance started</u>   Renaissance started in Italy at the beginning of the fifteenth century. We generally took Brunelleschi as the first interpretor of architecture.

It started in Italy because:

(1) <u>Social and political conditions.</u>

(2) <u>Classical Influence</u>

    (a) Persistance of classical traditions,

    (b) Presence of classical monuments,

    (c) Revival of classical literature.

(3) <u>Inability of the Italians to accept Gothic architecture.</u>

-3-

Lecture 2, September 29, 1925.

Field of Study — The field in the study of Renaissance art is inexhaustible because we have more monuments. The documents are full. All arts well preserved, which offers us an opportunity to compare painting, sculpture, and architecture.

Renaissance & Modern Ages — There is a certain sympathy between our age and the Renaissance age: they are both commercial ages. For example, Medici, the great patron of art, are bankers.

We had no upheaval since the Renaissance — there are only small uprisings, such as Victorian Gothic in England, the Greek Revival, etc. Modern architecture is more influenced by Renaissance than any other period. We still do classical architecture.

Reasons that It started in Italy — In the 8th century, Pepin th Short, (752 assumed crown) king of France, conquered Revanna, and presented it to the Pope. Charlamagne, Charles the Great, was Pepin's son, planned to unite the Roman Empire. He finally conquered Lombardy, which was in control of Italy. In 800, Charlamagne was crowned in Rome by the Pope as King of France and Emperor of Rome.

-4-

He held title some what dependant of th e Pope. Power of Pope continue to increase in a practical and spiritual sense. Rome being the center of government, men naturally turned to it.

<u>German Emperors</u>   At Charlamagne's death; the Empire was divided by his sons. In th tenth century, Otto the Great, went down to Italy, crowned at Milan and Rome, and founded the Holy Roman Empire. The German Emperors were very jealous of their titles, and spent much energy to maintain it.

<u>Conflict btwn Pope & Emperor</u>   At the same time there was the conflict between the Pope and the Emperor. Two parties rose, the Gueles sided with the Pope and the Ghibellines sided with the Emperor.

In 11th century, Pope Gregory VII went so far as to claim power whether the Emperor be crowned or not.

<u>Feudal Age</u>   All over Europe, gvernments were very unsettled. In the north, feudal system was in vogue, and well established and strong. Fidels got protection from influential lords. Every body expected to figh when called upon. No peace but in

monasteries.   Nobles very strong.

William the Conqueror made every body owe ligion to King, at the same time, French king was doing the same hing.   Units formed by lords and followers were absorbed into one nation.

<u>Feudal system</u>   In Italy, feudal system was not a
<u>Failed in Italy</u>   success.   Communes or cities
were the fundamental ideas.   Copporations of artisans and guilds formed city.   Each guild formed a complete unit by itself.   In Lombardy, church unite with people against king.   And in   , people unite with nobles agaisnt th Church.   In 11th century, gradually came back to city again.

The Church and tis representatives wee more national than Emperor.   But city never wholly accepted Pope or Emperor.   The cities were playing one against the other.   They formed independant republics.   The propietors, German nobles, really fought to live in the city, and gave truubles to the people.   In 13th century the cities managed to find the bandit tyrants outside of city.

In 1250, German bobles gave up claim for The Italian Peninsula.   Pope moved b Fance, so Italy was left without ruler.   Cities left in undisputed control.

<u>City &</u>
<u>Nation</u>
The administration of these cities had been a strong factor in the education of the people. Individuals were forced to submit entirely to power of the public. Entrance into society sharpened their wits, but very few entered into society. Among this minority they produed diplomats and statesmen.

In 1850, at the unification fo Italy, one of the great difficulties is the alligiance of people to city. They did not ally with the nation.

Art of this period largely communal.

Unity between cities entirely lost. Florence had to get Pisa to get to th sea, and get Sienna to control road to Rome. When one city wasconquered, there was no attempt on the part of the conqueror to mix up with the conquered. They would make them as uncomfortable as they could.

During the 14th century, certain families got control of city, such as the Medicis. They would hold the position by any means they could find. They were a sort of glorified political bosses. There was constantly the danger that some one will get upper hand and overturn them. Alliance with the intellectual class was the most honorable step this kind of government had taken - they gave them dignity. As these tyrants all had the desire to leave behind them monuments bearing their names, and they had large sums at their disposal, a good deal of the energy was turned into he channel of art.

①

Lecture 3 October 5, 1925.

<u>Condottiere</u>   A later phase of tyranny was those founded by the "condottiere" – military captains brought in by tyrants.   Citizens were glad to be reliefed of fighting.   Captains could be perchased, and he brought his men along.   Tyrants were glad to do this also, because they were paid and would not turn against him. But the condottiers, realising their own power, rose and overthrew tyrants.

<u>Development of Individuals</u>   The principal interest which all these have for us is to show the ~~development of indivicuals~~.   This is <u>a strong factor</u> of this revival.   Self conscious freedom is due to system of government.   From beginning of Renaissance the individuality and talent of one man is going to influence architecture.   <u>Personality</u>, which the Middle ages tried to suppress, <u>began to liberate</u>.   In communal guild, city, tyranny, and condotiereship, all held position by force of their own talent.

(10)

<u>Survival of
Latin Tongue</u>　　　The middle of the 13th Century and the beginning of the 14th brought in 3 great men in literature: Dante, Petrach, and Boccacio - all had great interest in classic literature. If art should have any revolutionary movement, lieterature should feel it first.

Although Dante is spoken of as the last of the middle ages, yet he is in the threshold.　Petrach is the first of Renaissance writers.　All three raised interest of classic mythology and created Renaissance literature.　They wrote in the "vulgar tongue".

<u>Revival of
Classis Culture</u>　　Exterior of classic art is another strong influence of revival of classic culture.　All through the middle, they had these monuments. At the beginning of Renaissance were many standing which have since disappeared.　Some, during Renaissance, were destroyed to be incorparated into Renaissance buildings - Colleseum is a case. Some of the marble went to lime makers.

Cpature of Constantinople in 1453 brought in many scolars.　At the beginning of 13th century there had been a sort of classic influence in work of Michael, Pisano, a sculptor.　He had certainly turned to study of relief in Roman sacorphagus in Pisa and Siena. Although his work is in Gothic, there showed classic influence.　Through Gothic period there were ornaments suggesting classic influence. In painted ornaments also, such as acathus leaves, etc.

Some writers on architecture and art think if Italy lived withough influence of foreign missionaries, they would have returned to classics sooner.

<u>Inability of Adapting Gothic</u> Gothic is not native to the Italians. <u>The Structural problem did not interest them.</u> In stead of using buttresses, they prefered tie rods, although not very monumental. They still use wooden roof of Basilicca. <u>Climate make large area of glass undesirable.</u> Window area does not occupy 1/3 of space between columns. This gave large unbroken wall area. And it was intended to cover inside of this with paintings, breaking into small panels. A great advantage of painting is to give scale. Small panels give standard of scale. Outside wall is very impressive.

<u>Openness of Plan</u> Another characteristic which show influence of classical architecture is <u>openness of plan.</u> In north, side aisle is definitely separated. In Italian Gothic, side aisle and nave are like one thing. Height is about same in side aisle and nave.

<u>Local Influences</u> In Norht Italy more German influence. In South, Norman. Etrucsan influence in Tuscany.

Lecture 4, October 6, 1925.

Other sources of Study A chief source is bussiness account in cities. Another is Vasari, contemporary of Michael Angelo, had stories of artists, philosopher, poets, etc. of Early Renaissance. Not absolutely reliable.

Early Renaissance Early Renaissance or Formative Period, 1420-1490, roughly in the 15th century. Characteristics is that <u>inspiration drawn from Roman monuments and ornaments</u>. <u>Original in type and composition</u>. <u>Great freedom and grace</u> in composition, especially in North Italy. <u>Less archaicological</u> than later because they knew little about Roman buildings.

High Renaissance High Renaissance, the <u>golden age of arts</u>. Produced remarkable group of men. Michael Angelo, Raphael, etc., belong to it. First part of 16th century. Marked by <u>greater care in reproduction of classic ornaments</u>. Orders used in most of composition. Tendency towards <u>greater vigour</u> and less refinement. Modelling heavier, with heavy shadow.

(13)

<u>Formally</u>　　　　　　Latter half of 16th century.
<u>Classic Period</u>
　　　　　　　　　　　Period of Paladio and Vignola.

A period of <u>Mathematical working out</u> of architecture.

Characterised by devotion to proportion of orders.

Collosal orders ~~to~~ were also characteristic.

<u>Rococo or</u>　　　17th centhry.　　Marked <u>reaction</u>
<u>Baroque</u>
　　　　　　　　<u>against formulaesm</u>.　　Used stucco,

sometimes paited to indicate marble.　<u>Free use of fig-</u>
<u>ures</u>.　<u>Bold and coarse ornaments</u>, because of stuuco.

Free curves, because it is easier than straight line

for stucco.　Tried orders to indignities.　Broken

entablature, etc., cap up side down.

<u>Classic</u>　　　　18th century.　　Became <u>simple</u>, clas-
<u>Reaction</u>
　　　　　　　　　　sical, and pure.

<u>Early Renaissance</u>

　　Had three centres of developments, may be said to

have tree school.

1 <u>Florentine</u>
<u>or Tuscan</u>　　　Center at <u>Florence</u>.　　Had wide in-
fluence.　Include all of Tuscany, *Umbria*, even

Rome and Italy.　May be said to be school of every-

(14)

thing south of Florence. Stone was main material. Design characterised by purity and symmetry. Depend mainly on proportion and conception of problem. Müntz French writer, characterised it as "School of Purist".

2 Lombard School Great plain of Lombardy, Northern part of Penisula, little to the West.

Center was Milan. Include Pavia, Bolgna. Little stone produced. Great deal of brick and terra-cotta used. That led to multiplication of ornament, because using of models does not increase labour. Ornate sort of stybe. Retained good deal of Picturesqueness of Gothic. Müntz called it "School of Fantasies". *Fantasies.*

3 Venician School North-east of Pennisula, Venice as centre. Venice was in contact with east. Byzantine tradition very strong. Use of marble for incrustation was characteristic. Use of colour shows Oriental character. Münts - "School of Colorists".

In Renaissance, individuality of artists is so important that we have to study their biographies.

(15)

FLORENTINE SCHOOL

<u>Brunelleschi</u>   Filippo di ser Brunellesco or dei Brunelleschi is first great figure. 1377-1446. Father was lawyer and was given tittle of Messer. Filippo followed his inclination and was <u>apprentice to gold smith</u>, and spent time in <u>mechanics</u>. Gold smith in Venice was famous. Most architects got trainin in gold smith's work. They got training of drawing and of sculpture there. Brunelleschi got interested in sculpture and got in touch with Bonatello. In 1401, compitition held for design of Baptistry doors. Trial consisted a panel representing sacrifice of Isaac. Compitition went to Brunelleschi and Lorenzo Ghiberti. Jury suggested they worke together. According to Vasari, it is said that Brunelleschi generously resigned in favour of Ghiberti. According to some contemporary historian, he refused. The latter seem more likely true.

Went to Rome with Domenichino, study classic ornaments. Worked in gold smiths shop to earn living.
H He had two ambitions, (1) To bring back the architecture of Rome. (2) to complete dome of Cathedral of Florence which was begun by Analfo di cambo in 13th Century.

(16)

When Brunelleschi returned to Florence, the dome had been brought to a part above the circular windows of the drum.

(18)

Lecture 5, October 12, 1925.

<u>Dome of Ca-
thedral of
Florence</u>
Plan of Dome was <u>octagonal</u>, with <u>four</u> large piers alternating with <u>four</u> large pointed arches.

Brunelleschi was called in by those who were in authority, but they could not bring to any decision. He left Florence suddenly in anger and they begged him to return. They wanted him to make some exposition of scheme. He refused. He suggested that a congress of the world's architects be called to a discussion. This was held in 1420. In fact only Italians took part. The template of the construction was the greatest question. One scheme suggested that a mount of earth should be constructed, mixed with pennies, to form as a sort of centering. And some similarly foolish schemes. His scheme was so radical that they chased him from meeting. But he went on and won over favour of members of guild, by converting each separately. At last he submit detail of his scheme.

(19)

Although he had all the scheme, he himself did not know exactly what was going to happen. He proposed to construct an <u>octagonal dome in double shells</u>, with<u>out centreing</u>. They finally accepted. But forced him the service of Ghiberti. This kept his hand tied and situation made difficult. In 1423, Brunelleschi made model &r chain of dome. 1425, Ghiberti's salary discontinued. Continued again next January, but finally stopped in 1432. 1433, Brunelleschi became sole architect of dome.

Great octagon was not original plan. In 1348, Great plague in Florence, people fled from city and left money behind. Church enriched. Dome of p Pantheion in Rome is 140' in diameter. While he did not get inspiration from Panthion, he knew it could be done.

<u>Plan</u> He made dome in <u>two shells</u>. He decided to use in section a <u>high pointed arch</u> and decrease the thrat. The two shells wee connected with ri bs. <u>Eight</u> large one at the corners and sixteen small ones in between. These ribs constrated at same time as shll is build and <u>not a skelton</u> as Gothic. Besides rib,

(20)

construction was reenforced by horizontal arches.
Then a great bond of metal and timber, called a chain,
same principle as tie rod.   Lower part all in stone,
upper part stone and brick.   Staircases carried up in
between two shells.   Construction is not copy of
Roman but is result of his study of Roman monuments.
Theoretically, the dome is a cloistered vault on an
octagonal plan.

<u>Effect of dome on Architecture</u>   Effect of dome on architecture of the period is tremendous.

It is forerunner of dome of St. Peter's, and has a
tremendous educational effect.   All these buildings
was brought through the middle ages.   Ornaments
showed some classical influence, but it is the lantern
on top that shows real classic ornament.   Build without centering. Used spiral bond.

<u>Interior</u>   Brunelleschi intended to cover the interior of dome with mosaic.   But it was never done.

Left to hands of artisans.   And it was a fatal mistke
to have horizontal bands of figures.   It would be better
to have ribs accented.

(21)

Lecture 6, 15, October, 1925.

<u>Other Achievements</u>   Construction of dome is ¢
<u>of Brunelleschi</u>
                career of Brunelleschi's

life.   But he did som other.

<u>Pazzi Chapel</u>   Chapel of Pazzi family.   Date not

Certain.   Some said as late as 1430, but not likely

as some said as early as 1400.   Probably after he

went to Rome.   In cloister of St. Croce at Florence.

Free handling of design.   But cloister hinders porch.

Nevertheless, it is freer than

where the substructure is already

finished.   <u>Pendentive with bar-</u>

<u>rel vaults on both sides.</u>   <u>Vaulting of interior re</u>-

peated on outside.   <u>Columns Corinthian.</u>   Capitals

crude and not harmonising with othr parts of building.

According to some authorities, cap is form some old

Roman building.

This building displayed him as a decorator more than

any othr.   The dome is more like Gothic.   Has sirés

of ribs.   In between each of these are vaulted webs,

with windows comming below.

(22)

Interior is also decorated with Corinthian order.
Door simple and classical with pediment.

(24)

Lecture 7, October 20, 1925.

<u>New Manner of Building</u> Brunelleschi inaugurated new manner of building for churches.   Plan not much altered, except that the support is lighter. Litegy of the church remains the same.

<u>S. Lorenzo</u>   In S. Lorenzo, he used plan similar to type used in S. Croce (T Plan).   Not much projection of choir, trancept on either side of choir. Three aisles divided by archades.   Series of very shallow chapels open out into side aisle.   Nave and trancept covered with flat ceiling.   At crossing is pendentive dome.   Base of side aisle vaulted with pendentive vault.   Used Corinthian order, his favorite order.   Pendentive arches under dome supported by pylasters.   Pylasters are tall in proportion, because at corners gets diagonal width.   They are 13 or 14 diameters in height.   Used entablature block between arcade and column.   Facade never finished. Design was made for it by Michelangelo, but was never carried out.

(25)

Lecture 8, October 26, 1925.

<u>Loggia of Ospedale degli Innocente 1419</u>   Hospital of the Innocence. <u>Facade</u> treated with <u>long loggia</u>. Loggia treated with Florentine arcade. Terminated at either ends with some treatment like end pavilion, rather unsual for this early period.

<u>Other Buildings</u>   Brunelleschi was not confined to ecclesiastical architecture alone, he did two palaces.

<u>(1) Pitti Pallazzo</u>   Built forbrich Florentine Luca Pitte. Pitti was a rich politician who got his money not very scrupulouly. Had many enemies. Built palace with court to put enemies.

<u>Facade 400' long</u>. Originally only central three arches were built. Pitti soon lost power and palace fell into hands of the Medicis. Now is royal palace at Florence.

Brunelleschi's inspiration for this building is Etruscan monuments. Used <u>large type of rus-</u>

(25)

S. Spirito   315 feet long.   More regular plan: Latin cross with long front arm.   Side aisle carried around.   Florentine arcade with column.   Flat ceiling with vault in crossing.   This is unpleasant, but it makes rich interior. This shows he is somewhat inspired by Early Christian Basilicas.

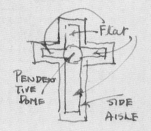

Monastery at   Plan, Latin cross.   No side aisle. Chapel opening out to nave.   Choir same width as nave.   Interior beautifully proportioned. Considered his master piece.   This chapel replaced an older building that has old Romanesque facade. Brunelleschi incorparated it into building, but never finished.

BV = BARREL VAULT.
PD = PENDENTIVE DOME.

tication. Great massy rock with arcading treatment for both doors and windows.

It is believed that even the central portion is not finished according to Brunelleschi's design. Had no heavy cornice. Some critics believed that it was intended to have another story with heavy cornice.

The plan was extended later. Now it has a shape like this:

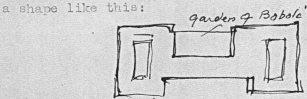

Garden extended on both sides up hills. The man who completed this palace is Amanati.

At present there are series of windows on the lower story. But all added later. Facade now 470' long. Base 24' high. Some rustications project as much as 2'.

It stands between Medieval fortresses and modern town houses. Still retain character of military building.

(27)

Brunelleschi prepared a palace for medici. When he showed it to Cosmo Medici, he said it is too elaborate for a citizen of a republic, add Brunelleschi destroyed the model in anger.

(2) Palazzo
<u>Quaratesi</u>   He also build palace for the Pazzi family.   Stone for the fist story, above is stucco.   Over-hanging roof of 3' or 4' in projection.

<u>As a Character in
History of Architecture</u>   He is a most interesting character in the history of architecture.   He is spoken as the strongest personal influence.   Thoroughly in view with humanistic and classical movement. Luckily the period has some personage who can express it so completely.

Müntz says he he has combined the precise mind of an engineer and th rifinement of an artist and the strength of a man of affairs.

In S Lorenzo ad others he contributed the new type of building

In Pazzi Chapel he used refinement of Romans.

Solution of problem <u>direct and frank</u>.   Most of his work depend solely on proportion and general

(28)

conception. This is true inspite of fact that someone says it is superficial and merely Roman adaptation.

Michelozzi   Micholozzo Michelozzi 1396(?)-1472.

Michelozzi is one of group of students around Brunelleschi. Native of Florence. First learned sculpture and engraving of seal. Working under Ghiberti. Friend of Cosmo Medici. Followed him in exile to Venice.

Master piece is palace bilt for Medici, now known as Palazzo Riccardi (1430). Mechelozzi's design was accpeted after Brunelleschi has broken his model.

The original is about half th size of what is standing now. Three doorways and ten windows, all the rest are added later. Built around a court, treated with Florentine arcade. Exterior wall rests on projecting base. Openings are arched on all the stories. Originally all

(29)

openings were doors, some filled in with windows designed by Michelangelo.　Stories devided by string course which acts as sills for windows. First story heavily rusticatedblike Pitti palace. Second , less and third Smooth.　The whole structure is crwowned by ah 8' projected cornice whose general character is like a Corinthian Cornice.

(30)

Lecture 9, October 27, 1925.

Other Works of
Michelozzi

<u>Convento of S. Marco</u>   In Florence.   Cloister is charming, but not a building of distinction.   Fresco interesting.

<u>Palazzo Vacquo</u>   Alteration of Palazzo, more engineering than architectural design.

Portinari chapel at <u>S. Eustorgio</u>   In Milan, Portinari family were representative of the Medici Bank.   One of women has devotion of S. Peter Marto.   She decided to build this to house the Gothic tomb at S. Eustorgio.   In this building has many character of Lombard school, — buttresses, terminated in Gothic tabanacles with Renaissance detail.   Interior scheme is like central unit of Pazzi chapel.   Pendentive dome openig chapel, head in frieze, etc.

(31)

<u>Alberti</u>   Leon Battista Alberti 1404-1472.   Born in Venice during exile of his family along with the Medicis.   Illigitimate son, but made no difference of his advantages, brought up same as the ligitimate children of the family and redeived same education. Very vercital type of man.   Went to university of Bologna, iinterested in sciehce, philosophy, etc. Latin scholar, published work in Latin, and also Latin poems.   Also first wrks of sociology.   Also and athelete.

1428, returned to Florence and Medici bedame his patron.   Contemporary of Brunellesch and Bonnetelo.

He is most scholarly person who devoted his life to new movement of architecture.   His adapting architecture as his profession raised the standard of architecture.   He displays a good deal.   More interested in design than in execution, which sometimes led to defects of wok.

We first hear him at Rimini on eastern coast of Italy, onga

(32)

<u>S. Francesco</u>   We first hear him at Rimini on the eastern coast of Italy, egaged in the <u>remodelling of the church of S. Francesco</u>, for Sigismodo Malatesta. Malatesta is one of the paid captain, very unscrupulus, morally guilty of many unspeakable crimes, but devoted to classic tradition, and has apprediation fo beautiful things.   He is surrounded by poets, and artists. He bought a body of a Greek philosopher from Greece to be burried in this church.   Changed to heathen temple and <u>dedicated to his mistress Isota,</u> whom he later married.

Alberti was forced to <u>follow to some extedn the line</u> of the old building in the interior.   For exteior, Alberti <u>made a new sheath</u>, even not car ing whether the opening come to line.   Shell about one meter away from wall.

Facade takes <u>inspiration from a Roman archi</u> in Rimini. Three arches in fromt, framed by <u>engaged columns</u>. This is first example of engaged columns used.   Above <u>central bay is incomplete smaller order.</u>   Sides <u>treated with deeply recessed arches on piers.</u>   Recesses are for spcophagas.   It is believed to be domed, but no trace in plan.

(33)

Interior has relief by Agustino di Duccio, flat and delicate. Architecture of interior not very pleasing. Müntz saied it is so bad that Alberti could not have done it. He had design it one way and might be altered by othrs.

Palazzo Rucellai 1451 In Florence. Work carried out by Rosellino. First palace facade in which orders are used. Used as slightly projecting pylasters, not even ½ diameter. Tuscan and Doric on lower floor, and Corinthian on top. Difficulty comes in proportion as is always the case with superposed orders. Alberti designed an entablature little heavy top, with brackets on frieze, making cornice, frieze and architrabe as one crowning unit.

S. Maria Novella Facade 1456. Built for Rucellai in Florence. Took inspiration from Romanesque. Colour marbel. He used for first time the big scroll to fill angle between high wall at end of nave and low wall at eng of side aisle.

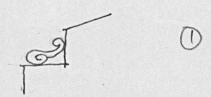

Lecture 10, November 1, 1925.

S. Andrea   1512.   Completed after death of Alberti.   Without side aisle, with chapels opening out into the nave.   Barrel vault over nave and pendentive dome at crossing.   Latin cross in plan. The chapels alternate with large and small.   The opening of the small one is so designed that it looks more like a pier than a chapel, and the effect of an arch and a pier is thus carried out.   Chapels covered with barrel vaults with axis perpendicular to nave, but below spring line and do not penetrate.   It is said to be inspiration for Bromanti when he designed S Peters.   Facade form Triumphal arch.   Tall arch in middle, flank by two piers, decorated with windows, whole thing surmounted by pediment with small features above.

Rosselino 1409-1464

He is connected with Alberti as Michelozzi is to Brunelleschi.   In some cases as assistant.   Ros-

(36)

selino and Alberti were both associated on some work on S. Peter, begun by Pidiolus V, started by laying foundation for apse. It is an attempt to remodel the Early Christian basilica.

<u>Pallazzo Piccolomini</u>   At Siena, completed 1460. For Pope Pius II, who is an interesting figure in the Humannist movement. In design for this Palace is great influence by Palazzo Ricardi. It nevertheless remain a great contrast to Ricardi. Extended frieze under cornice, emphasising crowning member without getting too heavy.

<u>At Pienza</u>   Also employed by Pope Pius II for a group of buildings at Pienza. Pope interested in the ancients. He like to name a twon after himself. Changed Corignano to Pienza after Pio. Palace is an literary copy of Palazzo Rucellai in design.

LECTURE 11, NOVEMBER 2, 1925

<u>Giuliano da San Gallo</u>

<u>S. Maria delle Carceri (Prato)</u>   The Florentine School of the First period, at its end a type of church appeared with Greek cross in plan, giving symmetrical on axis of both arms. Earlier one is

(37)

S. Maria delle Carceri.   Arms of cross covered with barrel vault.   Crossing by dome raised slightly by a drum.   It is not an invention, it is Pazzi Chapel on two axis.

Cruciform Plan is also used by Alberti.   At S. Sebastian at Mantua.   Exterior two stories and 2 Pylasters at corners.

Benedetto da Majano and Cronaca

<u>Palazzo Strozzi</u>   At Florence, 1489.   Finished 1 53, original scheme carried out.   Follow general scheme of Ricardi, but cornice is not so heavy.   Has 1 opening on each facade and small windows on first story.   Horizontal divesions make sill for windows. Rustication graded from top to bottom.   Court with Florentine arcade.   Still has logia on uper story.

Agustino di Duccio

<u>Oratory of S. Barnardino</u>   At Perugia, examples of a type of facade which owe interest to sculpture.   The architecture is but a frame. Building is used as a con-fraternity, for organizations of somewhoat religious natue.   It is a little chapel for meeting of those organizations.   Emblem is

Story is that S. Barnardino was preaching against gambling and man say he ruins his business, so he drew this and gave

(38)

the man fight to prit.   It is often found on houses
in Tuscany.   This facade is now charming, blue glaze
is splitted and showing pinkish orange terracotta,
better than when it was new.

<u>A Minor Production</u>　　　　　A class of work between
<u>of the Period</u>
　　　　　　　　　　　　　　architecture and sculpture,
such as pulpit, alter, lanterns, etc.   They are
particularly successful in this period.   Very elaborately
ornamented, used all Greek and Roman ornament,
but more delicate.   Has omething of Greek refinement.
Many people think of these when referred to Italian
Renaissance.

(40)

Lecture 12, November 9, 1925.

<u>Influence of Florentine School</u> Florentine School, being the center of the movement, had influence on other places. Rome had hardly recovered from the absence of Papacy. During this time, what is done in Rome is by Florentine architects who went down to thee. First Pope of the Renaissance is Nicholas V. He called Alberti and Rossellini to Rome to consider the alternation of S. Peter's although the plan was not carried out.

<u>Pontelli</u>

<u>Palazzo Venezia</u> Exterior is Gothic, crowned by buttresses, project out, supported by corbel. Door shows influence of gold smith's work. Architect really unknown, but is assigned to Pontelli. Court treated with Roman arcade. First Rwnaissance structure that employed that type of arcade. Order supported on pedastals. Court unfinished. By side of this is a small palace. Has been removed because it came into way of new Victor Emmanual Monument.

(41)

Summary

<u>General Conception</u>   Architect often not entirely free in design.   Several of the monuments are continuation of Gothic structures.   Simplicity of comception.   Flatness of treatment, slight projection.   Great restrain from treatment of ornament.   Fondness of Corinthian; Brunelleschi used it always.   For church facade, triumphal arch is probably inspiration.

<u>Use of Features</u>   <u>Florentine arcade</u> most popular.   Roman arcade also used - Rimini and P. Venezia, but not usual.   <u>Rustication</u> is another feature frequently found, it is one of effects revoked to.   Frequently graduated.   Rusticated arches. Astragus is a continuous line, radiating lines not tied into joints.   <u>Heavy cornice</u>.   <u>Doorway</u> has several treatments.   Round arch most popular. Square head door - Palazzo Rucellai.   Pediment - Pazzi Chapel.   Florentine Pediment also popular. <u>Windows</u>:   Most popular is twin window, it is a survival of Gothic, is near to plate tracery.

(42)

Square headed window - lower floor of P. Strozzi.

<u>Vaults</u>: Barrel most common.   Groined is type for courts.   Pendentive also much used.

(43)

LOMBARD SCHOOL

<u>Gothic Tradition</u>   In the north, Gothic tradition is stronger than in Tuscany. It is not easily set aside. Has tendency to retain picturesqueness of Gothic and elaboration of detail.

<u>Terracotta</u>   Another condition is the use of terra cotta to further this tendency. Lombardy is a large plain, little stone, but has good material for terracotta which is easy to reproduce ornament, easier than plain thing.

<u>Milan</u>   Center is Milan, chief town and capital of Lombardy. Portinari chapel in Milán by Micholezzi is disputed by Lombard writers. It has influence here.

<u>Filarete</u>   Antonio di Pietro Averlino, 1400-1469. He came in 1450. Entered into service of Fazza family. He worked in Cathedral.

<u>Ospedale Maggiore</u>   He determined plan and its treatment.
Scheme: Large court in center and 4 smaller court on either side.
Only small portion finished.

Exterior with Florentine arcade.   Later Solari came
and filled it in with pointed arches.
Red brick and terracotta, some stone
for columns.   But general colour is brick.

<u>Bramante   Donato di Angelo 1444-1514.</u>

Also belonging to the school are some works of
Bramante.   He belongs principally to the High Renaissance, but his early life was spent in Milan.

① <u>S. M. Presso S. Satiro</u>   His first work.   Has
some new features.   Perspective newly discovered
and Bramante, when he came to design this church, he
found that the apse will get into the street, so he
<u>did it in perspective in relief.</u>   He is <u>first one
to use bent order in corner.</u>

② <u>S. Maria</u>   East end of S. M. della Grazia.
<u>Della Grazia</u>   He covered it with <u>great dome</u>.
It is a Gothic church, but in it can be found influence
of Florentine dome.   Central space coming out are

(45)

covered with half conical roofs.

Used shaft instead of column.

Also buttresses and Medellions.

In terior used motive Brunelleschi
often used.

③ <u>Porch of Church</u>   It has influence on Bramante's
<u>at Abbiategrasso</u>   later works.   Arch on two

stories of orders. It is a Lombard porch except

with orders.

(47)

Lecture 13, November 16, 1925.

Character of Bramante's design completely changed after he went to Rome and working on different material.   Here he used terracotta, and it is interesting to see what material can influence character.

Certosa di Pavia	In town between Milan and Pavia. Facade of church and cloister very characteristic of Lombard school. Facade treated with buttresses. We see translation of tabanacle and pinacles into classic design. Facade in stone, but desing suggests much character of terracotta. Surface enterely covered wih ornament and bass relief. Scale is very small. Facade may be criticised that the ornament does not count when looked from some distance away. Certain parts inlaid with colour marble. It has character of a piece of jewel. Even somethings below eyelevel are so small that one has to knell down to see them. Too fine to be seen standing. Church itself is Gothic. Now a museum.

Cloisters are terracotta. Most of body of church is brick and terracotta. Facade never completed. There is representation inside of intended design. About half way up change takes place. Upper part more simple, but more an architecture.

⑤ <u>Doorway of Como Cathedral</u> — Gothic building. But doorway attributed to Bramante.

(49)

VENETIAN SCHOOL

Renaissance in Venice is 50 years behind Florentine. Last section in Italy to take up Movement. The reason is that because Gotic has firmer establishement. Also under influence of the East, continually receiving small objects of art from the Orient. Also under Old Byzantine influence. There are old Byzantine churches. Venice is modern ina sense: they did not have exactly same feeling for Roman civilisation as the rest of Italy. It has long been under Eastern Emperors.

Roman tradition largely eplaced by Byzantine. It is founded by people from main land to escape Barbarians, crowded on lagoons and built up city!

In 15th century these settlements grew till become a republic. Include Damacia and Crete. And then began conquest into Interior of Italy.

(30)

Lecture 14, November 17, 1925.

During first quarter of 15th century, Florence was allied to Venice against the Milanese, but later they became enemies. They were rather out of sympathy with this new movment. This rather keep them back. But through agency of Lombard artists, it is brought to Venice.

One of the picularities of this school is dynasty. For instance the family of Lombarde has generations devoted to architecture.

Fact in regards to personality on man is rare in the Venetian school.

<u>School of the Colourist</u>   Müntz called it the School of Colorists. It is most striking feature of schoo. Used Byzantine revecment. All over Italy they used same kind of construcition that the Romans used. Build a corpse and the architecture is a cover. Architecture is a revecment. But in Venice, it is different athat the revecment is frankly a revecment, without feeling as to structure. Marble used as wall paper.

From viewpoint of frankness, it is better than those in other parts of Italy.

<u>Horizontal Accenting</u>   Accenting of Horizontal is another feature.   Usually consists of entire entablature. Some used a bouble frieze, etc. to make it heavier.

<u>Other Features</u>   Semicircular pediment much used.
Some times a facade terminated in a pediment.   Perhaps come from facade of S. Marco.   Silhuette cuts sky.

<u>Ornaments</u>   Used ornaments inspired by sea. Sea horse, sea weed, shells, mermaid and also ships, ropes, Neptunes, etc.

<u>Porta della Carta</u>   In Venice.   By Bartoloneo Buon.
Gothic design.   Entrance to Doges Palace.   Has tendency to Renaissance. Crockets has little puti, shell for back of nich, all are classical.

(52)

<u>Court of</u>　　　　　First story circular arch.
<u>Dogeo Palace</u>
　　　　　　　Second floor pointed arches except arch for stairway on top is elliptical. Arch supported in Piers.　Inlaid, colour marble divide between floors, fully accented, full entablature.　Use type of window already seen Lombard window.　Semicircular framed in square.　Inspired by Porta Barsari at Verona.
In composition, the facade fails utterly.　Heavy piers over opening below.　It does not follow simplest rule of composition.　Good in decorative detail.　Giant staircase is exquisite.
<u>Pietro Lombarde.</u>
<u>S. ~~M. Dei Miracoli~~</u> *Palazzo Vandramini*　Most famous work of Lombarde.　Most logical in composition.　Semilar~~ities~~ in general scheme in Gothic and Byzantine palaces.　Two ends accented, central open. Government of Venice is more settled.　Most of the palace is on canal　It is a place not easy to start a riot.　Facade is open.　Lower floor is for

(52)

Lecture 15, November 23, 1925.

<u>Squola S Marco</u> These are schools or societies.
<u>Squola S Rocco</u>
S Marco is a conspicuous example of buildings terminating in semicircular pediments.   Devided into two parts, having quite different treatments.   Used perspective in two bays. 1 arcade and 1 caissons ceiling.   Look right only from one station point for each.

S. Marco [Roc] is not so arnate, more architectural. Famous for fresco of Tinoreco.

<u>Fa Giocondo</u>

<u>Palazzo del</u> In between Venice and Milan are
<u>Consiglio</u>
a number of towns affected from both sides.   Sometimes under dominion of one and sometimes other.   They show influence of both schools. At Verona is this palace.   It is by Giocondo, who was a monk, and later worked on S. Peter, with Raphael. It is a Conspicuous example of a Two-part composition.

storage.　　Very damp.　　　Principal room is above
entrance.　　Ends in solid panels.　　Whole thing
is charming.　　With frieze over accented.　　Window
filled with adaptation of tracery to classic.
Facade with superposed orders.　　Entablature at
to p.has heavy frieze.

<u>S. M. dei</u>　　　Scheme interesting.　　Unusual.
<u>Miracoli</u>
　　　　　　　No side aisles, just a wide nave.
Covered with wooden barrel fault.　　Sanctury is
on high platform, reached by fifteen steps, vaulted
with a pendentive dome.　　This raises ceremonials
practically on a stage.　　Many Lombard churches have
that feature.　　Walls covered with marble, laid
off in panels, different contrasting colours whoel
thing treated with definite schme.　　Detail exquisite.　　Some of best arabests are in this church.
Dome over sanctury is onion shaped.　　Facade divided
into two stories.　　Two tiers of Pylasters, upper
carrying arcade.　　Front terminated in
barrel fault, with same contoru as
barrel vault behind.

(53)

First story divided by pylasters, each bay then devided into 4 units. Almost a defiance of rule that not to use column on axis. First story open logia with arcade. Second with pylasters and windows covered by semicircular pediments. Windows filled in by two arch openings. In th North they like this kind of windows, they all come from a Roman monument here in Verona.

<u>Brescia</u> has three monuments.

<u>Palazzo Municipal</u>   Also known as <u>La Logia</u>. It is lower story of this that belongs to this period. Upper is added later. Uses Roman arcade – first example of its kind here in North.

<u>Monte dei Pietro</u>   Mount of Pity.

<u>S. M. Dei Miracoli</u>   Lower part of Facade only, upper is later. Extremely ornate. Many characteristics of Lombard school. Use Cantalogary shaft to decorate pylasters. Fine in scale. Semicircular pediment used as at S. Zaccaria.

(34)

Bologna

<u>Palazo Bevilacqua</u> Exterior rustracated. Has base. Heavy cornice at top. Here they use an inferior stone, many now disintegrated. It is case with this palace. Court is most interesting. Florentine arcade on first and also second. Upper bay is just one-half size of what is lower. Great difference in scale makes second story look light.

<u>Palazzo Fava</u> Has another characteristic of Bologna, the large brackets. Portico is characteristic of Bologna streets. Upper part supported by arcade over side walk.

(54)

Gambelo and Mauro Coducci

<u>Facade of</u> 5 stories.   Accenting hoŕzontal
<u>S.Zaccaria</u>
           bends.   Terminated in semicir-
cular pediments, with two semicircular curves.
An interesting feature on second tier.   Series of
arcade treated with shells, look like niches.   But
in plan is very flat.

HIGH RENAISSANCE

ROMAN SCHOOL

BRAMANTE - 1444-1514
- 1485   Palazzo Cancelleria, Rome
- 1502   Tempietto in S. Pietro in Montorio, Rome
- 1503   Great Court of Vatican, Rome
- 1506   S. Peters, Rome

RAPHAEL (Raphaello Santi) 1483-1520
- 1514   Employed on S. Peters, Rome
- 1516   Villa Madana, Rome
- 1520   Palazzo Pandelfini, Florence

BALDASSALE PERUZZI - 1481-1536
- 1506   Villa Farnesina (Chighi), Rome
- 1520-27   Employed on S. Peters, Rome
- 1532-36   Employed on S. Peters, Rome
- 1529   Palazzo Massimi, Rome
- 1527   Palazzo Pollini, Siena
- 1521   Palazzo Albergati, Bologna

ANTONIO DA SAN GALLO THE YOUNGER 1485-1546
- 1517   Palazzo Farnese, Rome

MICHAELO ANGELO BUONAROTTI - 1475-1564
- 1523   New Sacristy of S. Lorenzo, Florence
- 1524   Laurethian Library, Florence
- 1538   Design for Campidoglio (Capitoline Hill), Rome
- 1547   Employed on S. Peters, Rome

VENETIAN SCHOOL

MICHELE SANMICHELI - 1464-1559
- 1527   Palazzo Bovilacqua, Verona
- 1530   Plaazzo Pompei, Verona
- 1524   Porta del Palio, Verona
- 1557   Capella Pellegrini in Church of S. Bernardino, Verona
- 1549   Palazzo Grimani, Venice

SANSOVINO (JACOPO TATTI) - 1486-1570
- 1532   Palazzo Cornaro della Ca'Grande, Venice
- 1536   La Zecca (Mint) Venice
- 1536   Library of S. Mark, Venice
- 1540   The Logetta, Venice

FLORENTINE SCHOOL

ANTONIO DA SAN GALLO THE ELDER 1455-1534
- 1518   Madonna de San Biagio, Montepuciano

------------- ?
- 1508   S.M. della Consolazione, Todi

GUILIANO DA SAN GALLO 1445-1516
- 1490   Palazzo Gondi, Florence

RAPHAEL
- 1520   Palazzo Pandolfini, Florence

(55)

Lecture 16, November 24, 1925.

HIGH RENAISSANCE

More Uniformity   In this period, schools not definitely differentiated.   Never general, more uniformity.   Center at Rome, Papacy stronger. States of church became more important as political factor.   Rome was taking prominance of capital and attracted nobles.   Julius II, 1503-1513, was a strong influence in making Rome center of interest. He was strong, ambitious, and absolutely in sympathy with the movement.   Lio X, son of Lorenzo the Significant succeeded him 1513-1522, and did much to enforce it.

At 1492, death of Lorenzo, ~~Rome~~ Florence lost leader and replaced by no body.   1494, first invasion of Italy by French under Charles VI.   He claimed Kingdom of Naples, went all way through Italy.   It is this that hastened Renaissance in France.

(56)

<u>Principal Divisions</u> Naples, State of Church, Tuscany, Lombardy and Venice.   They were not strong enough to stand alone, and too jealous to unite.   Italy became in war under Francis IV and Charles V.   During these wars leagues after leagues were formed.   Lasted until 1529, when peace declared after sack of Rome.   Rome suffered very much.   Condition was terible.   All work stopped.   Yet The two places suffered least were Rome and Venice.   Although Rome suffered at sackage, it is not so often troubled as othr cities.   And Venice is isloated.   It is theses two cities that produced the greatest amount in this period.   As a result much energy went into military architecture.   Sanmicheli, for instance, shows his influence from this.

<u>Rome as Center</u> Rome is center.   Popes, Cardinals, Nobles, all undertook fast scheme to employ the talents.   In spite of fact of its leaders, Rome is not so comfurtable as Rlorence.

(57)

Leo was a Florentine, transplanting Tuscan architecture to Rome.   Real leader is Julius II, in harmony with the time.   Personally in touch with artists. commanded and see work carried out.

<u>Characteristics of Architecture</u>   Architecture marked by <u>increased Vigour in modelling, bolder projection</u>.   Facade more modeled.   Problem became <u>composition of mass than lines alone</u>.   In first period, they used parrellelgram on which they drew lines.   Now is question of combining different masses of various heights and forms.   For example, S Peter's.

<u>Education from Painters</u>   Painters were discovering perspective.   That might have to do to consder architecture as a series of plans. Instead of goldsmith's shop, being the entrance to the field of architecture, most are painters and received first training at atelier of painters.   Bramante and Raphael are <u>these</u> two prominant figures of its kind.

(58)

<u>Less De-</u>   Architecture depend less on decoration,
<u>coration</u>   more on architectural features.   More
study of Roman models, this might lead to loss of d
delicacy.   Study of ancient ruins continued, but not
any attempt to reproduce absolutely any Roman buildings.

<u>Continuation of</u>   In the former period, may work
<u>Unfinished Work</u>
             ~~was~~ to remodel the Gothic, this
~~(is)~~ disadvantage ~~ous~~ to the architect.   In this period
this condition did not exsist, but had to finish unfinished work of the former period.

<u>Standing</u>   Standing of architect well est-
<u>of Architect</u>
                ablished, due to general interest
in achitecture and also geniuses interested in architecture.

<u>Inheritance   of</u>   Inheritance of the middle ~~age~~
<u>the Middle Age</u>
                  was not lost.   Romanesque,
Byzantine, Gothic.   Such thing as balcony and lan-
terns were pinicales.   Even such ~~a~~ characteristic *things* as the
Florentine arcades were all inherited from the Middle
age.   For study, ~~study~~ drawing is not sufficient.

(59)

They supplemented it with wood models, some were work of the architects. Michaelangelo studied his architecture in clay.

<u>Featured</u> <u>Use of order</u>, in superposition, same tradition as in Colessium. Steadier below, lighter above. Corinthian which was favorite in first period in this period occurs less frequently. <u>Doric and Ionic more used.</u> <u>Cantalaugary shaft</u> disappeared, for simplicity. Proportion, Vetruvius was final. Colour still produced by combining diferent materials. Some walls painted in fresco in figure subject and later often with architecture in perspective, with all sort of archtectural details. Another means was <u>scrafito</u>. This method consists of mortar of two colours. First put light coat. Them cover it with dark coat, before dark coat is too dry, scratch and leave white coat out. Durable, some original is still existing. Process can be reversd.

(60)

<u>BRAMANTE</u>   Born near Urbino, little twon to the East, where Raphail came.   Studied painting and there is in Milan some of his paintings, frescos, and figure.   According to some biographers, he could neither write nor read, but he was clever enough to hide it in the court.   No record whether he visited Tuscany or not, buy some early works showed he had been there.

Established in Milan, 1472-1474, and did those works that we studied in the former period. 1497, Settled permanantly in Rome.   Rome seemed to exert great influence on him.   He threw off picturesqueness of Lombardy and assumed more dignified and restrained type of design.

(62)

Lecture 17, November 30, 1923.

<u>Bramante</u>   After coming to Rome, his design changed due to fact that Pope's court is more grand and more metropolitan.   Also due to material, here more stone is used.

<u>Pallazo</u>   Palace for Cardinal Riario, on a
<u>Concelleria</u>   regular plot.   Interior has court and church, as if one of the courts is roofed to form the church.   Exterior has no expression of church in interior. b At either ends of facade has <u>slightly projected pavilion</u>, rather significant in this early period. Lower story is treated with a basement, with Rustication, a modern way of attacking a problem.   2 tiers above are pylasters on pedestals, treated with ryhamatic spacing, 2 near together making a small bay, and than 2 further apart, making a wide bay, as in Mantua.   It is not a matter of coupled order.   Window is type common in the North.   Round arch framed in square head.   Bramante is credeted to have brought it down.

(63)

Courtyard is treated with Florentine arcade. He made corner a pier, ~~but~~ *not* necessary in construction, yet it satisfies the eye. For columns in interior he used antique, but not for capitals. Tuscan order used, but elongated. Upper story treated with pylasters, used brackets in frieze of top unit.

There has been much contraversy about the authorship of this building. Date of building is earlier than date he came to Rome. Ger*müller* made careful studies of Bramante and brings out the point that he sights a letter from somebody asking whether Bramante is in Rome and the date of that letter is before he settled there.

<u>Palazzo Gib*bard*</u>   It is also by Bramante, and it is a close copy of Concelleria. It is increditable that a man talented like Bramante should copy from one of his contempories.

<u>Tempietto in S. Pietro</u>   In one building Bramante designed, he seemed to do archaeology, but failed. Tempietto a small round building to mark spot where S. Peter was crucified. Much like Roman temple of Vista in plan, has colonade around it. Dome

(64)

raised on drum, high dome with lantern on top. That alone make it different from the Roman temple. Doric used used. Balustrades along base of dome.

He also designed a circular cloister, but never carried out.

<u>Palazzo Giro-toladia</u>   It has facade very similar to Concelleria. Not so successful in proportion.

Although some critics claimed that the excedution is better.

<u>S. Maria della Pace</u>   Roman arcade on first folor, with Pylaster and pedestals. On second a collonade on which two bays are equal to one in the lower. It is colonade not quite right. He made improvement one motive he brought down from the North. Used column over axis of lower, and used pier over lower piers. It gives trength and weak accent, and used contrast to give light effect.

<u>Lond colonade at Vatican</u>   These buildings brought him to prominance. Julis II, selected him to take work in Vatican. Built gallery to connect Vatican and Villa of Innocence VIII. With these galleries

(65)

to form a court of 1000 feet long.　　Villa high
and Vatican low, Bramante used brick stairway for the
two levels.　　At end of court is huge nich.　　These
galleries were begun, but Bramante died before either
of them were finished.　　Owing to haste of Pope, work
was badly built, and had to be taken down and rebuilt.
Inspiration for this is from order of Theater of Marcelus.

<u>St. Peter</u>　　St. Peter is begun by Alberti.　　It re-
mained unfinished, untill Julius II came to throne.
He gave Michaelangelo to design of his tomb, it is so
big that not any room in the old Basilica can put it.
So he decided to tear down and put up another, and Bramante
was selected to do this.

　　Begun in 1506, Greek cross in plan, with pendentive
at crossing.　　Angles of cross, filled by smaller domes,
on pendentives.　　Then outside of those are campaniles.
Plancame practically square.　　Scheme recalled a little
of Byzantine traditions.　　Pope was impatient and hastened
work.　　Defects appert in great arches supporting
dome.　　Some books said that defects were serious and

(66)

were increased in size.   Others said it is not.
Raphael started to plan heavier dome, and continued
on pier that Bramante had constructed.

Lecture 18, December 1, 1925.

RAPHAEL

Raphaello Santi, 1483-1520.   He really wasnot much more than an architectural designer.   He died at 36. Did much in other field of arts, but he was becoming more and more interested in architecture, would have done more if lived longer.   He was from Urbino.

S. Peter   He was architect of S. Peter at death of Bramante.   Revised plan of S. Peter to a Latin cross, illadvised.   Long arm in perspective cut off effect of dome, in spite of fact that it is finished like that.   He hardly has influence on S Peter. Much occupied with other matter.

Also commissioned to complete court of S. Damaso. He decorated the Logia, now known as the Logia of Raphael. Used arabesto, inspired by Roman paintings in bed of the Tiber.

Villa Madana   Situated outside of Porta della Paplo, and was completed by Romalo.   It is type of building not used as residence, but really for just an outing, a background for an outdoor fate.   Surrounded

(68)

by a large garden on sloped ground, slope laid out in terraces, fountains, It is near enought to Rome to get out in a short time, belong to a distinct type of buildings that the Renaissance had produced. It was destroyed in the sack of Rome, only part of it remains. Interesting to see use of stucco relief. It has a circular courtyard.

Palazzo Pandelfini　　In Florence Raphael designed the Palazzo Pandelfini, erected after his death. Under eitther Roman or Florentine school. In the design, he ratain heavy crowning cornice. Rustication around doorway and at corners. Opening treated with triangular or semicircular pediments, decorated with orders.

It is fair to call him an architectural designer. He had not had training in engineering, depends on o others for construction. He actually put throught very little.

PERRUZZI

Baldassare Peruzzi, 1481-1536, is a most interesting figure of his Renaissance. He is called the "Architect of Architects". Has great distinction about his work that appeals more to architects than to layman.

Born in Sienna, by Florentine parantage. He was a painter. In 1503, came to Rome, and became pupil of Bramante. Extremely modest, and this prevented him from becoming a public figure. Works in Rome, Bologna and Sienna.

<u>Villa Farnesina</u> Originally built for Agustus Chighi, Rich Sienese banker. In this building is also a building for entertainment. Has two projecting wings. Bold projections, orders, and arcades. Has wide frieze.

<u>S. Peter</u> He was employed to work on S. Peters, 1520-27, 1532-36, but did little, could not exert influence of his individuality. He returned to Greek cross - a brilliant plan.

<u>Palazzo Massimi, Rome</u> Built to replace residence of the Masimmi family. Through street where principal facade is located in a curved and narrow street. Plot irregular. It was to dwell

(70)

two families. 2 rooms for projection in front.
Over is logia.   This gives strong accent and deep shadow.

In courtyard, colonades only at two ends.   Plan has more quality found in French plans.   Has cleverness in treating corners and exis.   This plan shws that he is also a master of decoration.

Many profiles show he had studied Greek architectture.

His othr buildings are the Palazzo Pollini, Siena, and the Palazzo Slbergati, Bologna.

(72)

Lecture 19, December 7, 1925.

San Gallo the Younger was a Florentine. Pupil of Bramante and became familiar with the Problem of S. Peter. Four other members of his family were architects.

S. Peter   Appointed to work on S. Peter after Peruzzi. He had a plan and a model, but did not l leave any trace in the existing building. It was compromise between the Latin and Greek crosses. Uses open vestibule and another closed vestibule in front. Latin from exterior and Greek from interior.

Palazzo Farnese   This is the palace where his fame lies.
The upper story and cornice is by Michael Angelo. Returns to Florentine palace, without order, although he intended to use it on the top floor. Rustication and quoins at corners and openings. Facade a little monotonous. Makes impression through size. and scale. Window with circular and triangular pediments. Court is entered by a vestibule with barrel vault, perhaps suggested by corridor underneath seats of Collesium. Court has Roman arcade in 2

lower

(73)

lower stories, third has pylasters.   First story has
small rooms.   Important rooms on the second story.
Erected for Alexander Farnese when he was cardinal,
Later elected Pope ~~Alexander~~ III.

Lecture 20, December 8, 1925.

<u>Michael Angelo</u>　　　Florentine,　　Studied painting
<u>1475 - 1564</u>
　　　　　　　　　　and sculpture.　　Pupil of a school
which the Medici family started allowing students to
study their collection of antique statutes.　　He is
generally known as a sculptor, but his work in apaint-
ing and architecture also great.　　Reputation as a
sculptor was established when Bramante came to Rome.

<u>First</u>　　　First work in architecture is facade for
<u>Work</u>
　　　　　S. Lorenzo, but never carried out.

<u>New Sacristy of</u>　　　At Florence.　　In the tombs
<u>S. Lorenzo</u>
　　　　　　　　　of the Medici family, shows
character.　　Supressed every thing but architecture.
Greater freedon in treatment - broken cornices, etc.
Introducing some restlessness in design.　　Large consols and scrolls also used.

<u>Stairway in Vestibule</u>　　　At Florence.　　Not very
<u>Laurentian Library.</u>
　　　　　　　　　good architecture.

<u>Campi-</u>　　One of his first works in Rome.　　Greater
<u>doglio</u>
　　　　　Part executed after his death.　　The
<u>Senate</u> with the <u>Museum</u> one one side and the <u>Conservatory</u>

(75)

on the other. Approached by broad stairway. In all these used clolssal order.

He is given credit for decline of Renaissance and introducing Baroque.

Farnese Palace — Cornice by him. Won by unfair competition.

S. Peter. — Had his commission form five of the Popes 1547-1564. He was to receive no inmureration of the work.

First step he took is to return to Bramante's scheme of Greek cross. Simplified it and added to frunt a porch inspired by the Pantheon. Raised dome on drum. Made freehand curve approaching an ellipse. After 1549, no fund. Due to a plot by friends of San Gallo to get him out of the work. His attitude was pathetic. At his death, scheme is few feet above ground. Made model of dome, which was ultimately followed. In model dome has three shells, but lower shell was abandoned during his life time. Vignola, who was in charge of Michael Angelo's work and della Porta, has credit for small cupula on the right.

In 1585, when Pope Sixtus V came to the Papal throne, he provided for the removal of the Oblisk that stood on the Circus of Nelo, and put it in fron of the Church. Sixtus V also determined to complete dome according to Michael Angelo's scheme. From January 15, 1588 to December 17 1588 was time for the contruction of the drum. December 22 1588, May 1590 was for construction of dome. 600 skilled workmen worked day and night.

Architecture of Michael Angelo is severly criticised. But Garnier, architect of th e Paris Opera House, said the line of dome is born of genius. But One has to go round behind the building to see the true effect of the dome. During the time of Clemens VIII, lantern put one. Inteiror by Fontana.

East end that stood on old Roman circus is unsafe. Wall is leaning three feet.

T They used Latin cross because wanted to cover area of old Basilica.

Michael Angelo's was abandoned. Facade completed

(77)

by Maderna. He decorated the interior with revecment. Modeled the North fountain.

Bernini made Bronze canopy over the altar. Much opposed at first, bout completed after nine years. It has often been stated that the copper is fron the ceiling of Porch of the Panthion. He also built bell tower on left of facade as you face it. Decorated great pier unde pendentive, useing twisted columns used in the Old Basilica, supposed to be from Greece.

Bernini was charged for causing the settlement in the foundation of the bell tower. Exiled, but later returned to work on S. Peter.

1650, Jubilee.

Pope Alexander Commissioned him to construct the Collonade in front of the church. He used ellipse. for plan. Collonade formed by four rows of Tuscan columns.

Clemens X finished the fountain on the South.

Vertical axis of ellispse is 1100' long.

Lecture 21, December 15, 1925.

VENICE is another artistic center. Rome took lead, was most important. But Venice is next. Both Rome and Venice were two large cities not involved in wars with France and Charles V. Venice was out of the pass and isolated. Lombardy was effected most. Army went through Milan.

Micheli [Sanmichelli] Father and Uncles were architects. At 16, he was sent to Rome. Clemens VII appointed him with San Gallo to fortify boundaries of the Papal State. Great part of his work is millitary architecture. Design severe. After he completed his service of Court, he went to Venice. His principal work is in Verona.

Palazzo Bovilacqua Lower story is basement. Roman Arcade with pylasterw. Heavily rusticated. Traditionally he is inventor of rusticated order. Above, treatment changed. - Lighter, much sculptural decoration. Rhythmic spacing of orders. Great deal about upper story which will link him to early Lombard Period.

<u>Palazzo Pompei</u>	Much more severe. Lower story rusticated..
<u>Porta del Palio</u>	Most famous of his works. It is one of the gates of Verona, of which he did

several. In these he combined the middle age fortified city gate with the triumphal arch.

<u>Capella Pellegrini</u>   In the Church of S. Baradino. He showed that he is capable of delicacy. Circular chapel, covered with dome with lantern. Over alter are pediments which follow curve of the room.

(82)

Lecture 22, January 4, 1926.

SANSOVINO

His name is Jacopo Tatti. Takes Sansovino from his master. Went to Rome as sculptor and studies under Bramante. Sick, went back ro Florence. and competed with Michaelangelo in facade of S. Lorenzo. After sack of Rome , went to Venice, where his princilal works are.

<u>Palazzo Cornaro della Ca'Grande</u> High lower story, rusticated; above which are two stories of arcades with coupled engaged columns. Large frieze on crowning entablature with windows of top floor in it. Three part composition almost lost. Slight suggesstion of it at ends.

<u>La Zecca (Mint)</u> Rusticated orders. First time of that kind of order in cylinder. Column treated with large and small drums. → DIAMETER

<u>The Logetta</u> At base of Campanile of S. Marco. A work of sculpture. Architecture is merely a frame. Destroyed when Campanili fell, but fell

(83)

from it.   And it was gethered and put back.

<u>The Library of S. Mark</u>   Facade with two tiers of Roman arcades.   Lower Doric and Upper Ionic.   On upper tier the order is supported by small columns.   Entablature is excessively heavy - 1/2 height of columns.   Frieze has windows, treated with figures in high relief.

FLORENTINE SCHOOL

Florence lost its leadership and has few work in the sixteenth century.

Two interesting churches useing the Greek cross.

<u>Madonna de San Biagio</u>   Similar to church of S. M. Grace by San Gallo.   High Dome

<u>S. M. Della Consolazione</u>   Terminated arms of cross in half polygons.   Authorship unknown.

<u>Palazzo Gondé</u>   By Guiliano da San Gallo.   Follows Scheme of old Florentine Palaces.   But reveals are slighter.   Inovation is staircase in courtyard and carried bewteen columns of arcade.

<u>Palazzo Pondelfini</u>   Might be catalogued to either Tuscan or Florentine School.

(84)

Lecture 23, January 5, 1926.

FORMALLY CLASSIC PERIOD 1550-1600

Second half of the sixteenth century. Term to designate most characteristic of Palladio. There is tendency towards greater fidelity to Roman details. Measurement and drawings which were accumulated crystalized into a proportion. Rules drawn up. Orders used as basis of all design and all sorts of experiments were tried with them.

Two characteristics were directly tracable to in fluence of Michaelangelo.

- Supression of all but architectural details
- Employment of large orders.

Correct and difnified, not ugly but sever and dry. Lack inspiration and varieties. Lack individuality. Suffers from too much technique. As if rule become more importan than thig itself.

Excessife use of stucco chapes architecture. It makes building look less permanant. This led later to painting face of structure with architectural features. Also led painting of stucco to look like marble, etc.

VIGNOLA, GIACOMO BAROZZI DA. 1507-73

Known by name of his town, in North of Italy. Studied painting at Balogna, not successful. 1535 came to Rome and devoted to study of architecture. His fame lied in the five orders - published 1563. Now still a decided influence. But he does not often use his own rules. Refinement and elegance of his work showed him to have been a pupil of Perruzi.

<u>Villa</u> Earliest work is Villa of Pope Julius
<u>Papa Giulio</u> IV outside a gate of Rome. Desiged as a back ground for a day in county. This, like most of his works, has some original features. Here is a large court terminated in semi-circule. Large pavilion in front containing rooms. Garden in different lefels and courts, one of which has sunk canal and fountain. Another is an incline in spiral. Much beautiful detail and stucco decoration. Similar to some of the Roman tombs. Building is now Etruscan Museum.

San Andrea Rome — Near to Villa Papa Giulio is San Andrea. Elliptical dome on Pendentive. Profile is low. Front has pediment.

Villa at Caprarola — Most pretentious is Villa at Caprarola built for Alexander Farnese. Plan is pantagonal on exterior and built around circular court. Sloping ground surves as terraces. Has Military character and shows a little bit of French chateau.

Il Gesu — Home of the church of the Hesuit Order. Vignola is only accountable for the schme. He used same parti as S. Andrea at Mantua by Alberti. No side aisel. Latin Cross. Greater width of nave. Interior belong to the next period. Facade by della Porta.

Cupolas of S. Peter's — Vignola followed Michalangelo on owrk of S. Peter's. The small cupola at the right is his, and the other is excecuted later according to the one he built.

Vignola is frequently severely criticised for jimpeding the developement of architecture. This is unjust. His work shows considerable invention and originality. If he invented rules, he employed them with taste. Critics can only critise those who use them.

PALLADIO, AUDREA 1518-1580

Born in Vicenza, little known of his early life. Was a stone cutter, attacted the interest of patrons who sent him to Rome where he studied and published his famous book on architecture.

<u>Basilica at Vicenza</u>   In 1549 he constructed the arcaded gallery in the Mediaval Townhall now known as the Basilica. Bay is wider then he wanted for his classic architedcture. He supported his arcade on a smaller order and invented his famous Palladian Motive. Really it is a combination of Florentine and Roman arcades. It is evident that he is embarassed by the space hat he is given to construct so he made a smaller space for the turning arcade at the corars. Entablature break at each column forming a sort of buttress effect.

(89)

Lecture 24, January 11, 1926.

PALLADIO (Continued)

<u>Basilica at Vicenza</u> (Continued) Roof is curved like a cloister vault, but it is of wood. The building is in stone, not of stucco which is c characteristic of his work.

<u>Palazzo Porta</u> At. Vicenza. Rusticated basement, Flat arched window with semicircular reliefing arches. 2nd. story has engaged Ionic columns. Pediment windows, entablature broken. Attic around terminates facade.

<u>Palazzo Chiericati</u> Vicenza. Two stories of orders. Doric and Ionic. Open logia at either ends. Reclining figures on pediments of second story window tracable to Michaelangelo's Medici tomb.

<u>Palazzo Valmarana</u> Colonade. pylasters. Composite order. Surmounted by attic. Last pylaster at ends left out, figure in relief is used. Anderson even defends it.

Palazzo Porto . Ionic order on first floor and
Barbaran   Composite on the second. In
form of engaged columns. At corner the
turning is treated like this: *almost Baroque* ⌐ *cosa del Diavolo ?*

Loggia del   Not finished. Would be largest of
Capitano   his work if completed. Composite.

Villa   Or Rotunda. Just outside of Vicenza.
Capra   Square mass, to four sides of which are
applied Ionic porticoes. Surmounted by pediment.
Center is domed hall, exterior profile of dome is
low. Pushing out as hipped roof, ugly.

San Giogio   In Venice Palladio built two churches.
Maggiore   S. Giogio Maggiore is fine in proportion, but cold in colour, especially as compared
to others in Venice. He attempted to express section in facade. Classic facade. Two orders of
differat heights. Larger order raised on pedestal.
Latin cross in plan.

Il Reden-   Latin cross in plan. Has no side
tore   aisle. Orders of different height
on the same level. Not so successful as method
used in San Georgio Maggiore.

(91)

<u>Theatro Olympico</u>   Vicenza.   Built for academie Olympico Society of Vicenza, in which to give c classical plays.   Begun just before he died.   He used ellipse instead of a circle as basis of his scheme.
~~Rows~~ *is 4 seat* in tiers like Greek and Roman theatres.   Scene permanant, of wood and stucco.   Doors open to street, and architecture in it is built in perstpective.

Columns elliptical in plan

(92)

Lecture 25, January 12, 1926.

During this period Genoa became prominent in the production of Plalaces.

<u>Palazzo del Universita</u>   By Bianco.   Different levels gives opportunity to build in successive height.   The Problem produces inclined vista. - a triuph of architecture.   Exterior not very interesting.   Often not modeled escept cornice and doorway All other features painted - even with shadows and reflected ligh.

<u>S. Maria in Carignano</u>   By Alessi,   Plan like Bramante's scheme for S. Peter.

Florence is not so prominent in this period.   Its only mark is the Garden facade of Pitti Palace, by Ammaniti

BAROQUE 1600-1700

Baroque as compared to Formally Classic period is the other extreme of the scale.   Curved lines used.

(93)

Orders ignored. Twisted shafts of columns. Pediments of double curvature. Broken pediments. Pediments turned inside out. curved plan used in plan of elevation. Arches formed of broken curves. Window openings like like trifoils. Scrolls at corners over worked.

In figure sculpture drapery intrigate folds. Action dramatic. Everything intend to produce effect of restlessness. Figures not placed in architectural frames, often fall out of architecture. Scale of decoration heavry.

Stucco continued to be used. Some of these are due to the material - easier to use curve lines. Painting of plaster to imitate stone, ect., and gilding and painting with metalic paint was practiced. Most offensive esample of this type is in Jesuit Church. It is a style under severe criticism. There is reviving of Baroque in recent days.

MADERNA

<u>St. Peters</u> Most important. Changed plan to
<u>Rome</u> Latin cross by adding front bays.
He also did vestibule and facade.

Lecture 25, February 8, 1926.

FRENCH RENAISSANCE

Golden Age in France was period of Phillip Augustus, 1118-1223. Louis IX 1223-1270 succeeded him. Period of Crusade was a period of activity. *in Cathedral building* France had decided influence in Intellectual world in Europe. University of Paris was prominent.

After this is Hundred years War, 1337-1453. Depression. English in France. French King only nominal. Jean d'Arc terminated the period, with coronation Charles VII in 1422.

Louis XI, 1461-1483 was last of the Middle Age Kings. At the end of the Middle Ages was new sense of Nationality. King was firmly established, really head of government.

After the Hundred Year's War was revival of trade. Bourgeois came into prominence. Was revival of classic study.

<u>Gothic</u>
<u>Gone</u>   Gothic had abundoned its course. No new vitality. Problem it set itself is solved. Energy devoted in inguinuity, dexterity, intricate mouldings, etc. No longer met requirement of the civilization. Church lost power, military architecture no longer necessary. Chateaux *& Hotel* principal problems of th period.

(94)

BERNINI, LORENZO 1589-1680

<u>Collonade</u>
<u>S. Peters</u> — Not at all characteristic of the period except the figures on top are little dramatic.

<u>Scala Regia</u>
<u>Vatican Rome</u> — Long narrow stairway between two walls. The stair gets narrower as it goes up higher.

LONGHENA, BALDASSARE 1604-1675

<u>S. Maria Della Salute</u> — Built in commemoration of a plague. Most sucdessful work ß the period. Plan octagonal. Surmounted by dome. Sanctuary at back by smaller dome, and apses opening out from it covered with ½ domes. Use of small dome with a large one is not a very consistant thing to do, but it is well done here because it is on a very small piece of island and the rear is as clear as the front. Scrolls used as butresses to support dome criticised to be too heavy.

<u>Palazzo Pesaro</u> — Almost reproduction of schme of Capitanio palace. Used free standing columns.

<u>Palazzo Rezzonico</u> — Also suggest Sansevino.

(95)

EIGHTEENTH CENTURY

Reaction against Baroque. Practically Palladien went back to formal. Not ugly, good proportion, but lack inspiration. Hard and mechanical.

Palace at Caserta By van Vitelli. Near Naples.
Square plan, 800' on each side. Divided into 4 courts. Galleries meet in centre where is an octagonal feature with grand staircase. Plan is often a crip for school.

Braccio Nuovo At Vatican. By Stern. Is used as sculpture gallery. Interesting to see the way that schulpture is incorporated with architecture.

(95)

<u>Italian Inspiration</u>   Inspiration for new life came from Italy.

① Had been in various contacts with Italy.

② + Holy Sea

First by travellers - rexlation with Vatican had taken eclesiastics to Italy.   Cardinal Ambois was great patron.

③ Embassy to Italian courts and Vatican.   Merchants and bankers.

④ Importation of small peices of art, pictures, furniture, metal and earth works, etc.

<u>Campaigns In Italy</u>   ⑤ The most important was compaigns of French in Italy.   Charles VIII claimed crown of Naples. Went to Italy in 1495.   Passed nothern to Naples.   Renaissance in Itlay was then well underway.   Louis XII added claims to Milan, 1499-1504 was his campaign, and continued later by Francis I and Henry II. 1559   These piople having seen Italian buildings in a new way, the noble then stared to do the same.

⑥ <u>Marriages</u>   Marriages of some of the French Kings to I Italian princesses was another reason.

Two princesses of the Medicci, Katherine to Henry II, and Maria to Louis XIV.   Luxamburg was designed after the Petti palace to remind the queen.

⑦ <u>Students</u>   In he reign of Louis XIV, French artists went to stdy in Italy.   French academy in Rome founded.   Italian artists brought to France.

(96)

<u>Contrast</u>　　　① French had litle first hand knowledge of classic architecture.　　Little remains in France.　　They got inspirations throgh Italians.　　② System of planning entirely different.　　French expresses every element – orderly disorder. Each part forms a unit.　③ Stair cases often spiral in round or polygonal　④ Roof high and steep, this brought along chimneys. High roof gave space underneath it and dormer window came.
⑤ Unsymmetrical.　⑥ Good Silhouette.
Italian opposite, symmetrical, all under one roof.　　Pitch of roof low, low chimney, no dormer.　　No expression of internal arrangement on exterior.

<u>Contrast in Construction</u>　　France used Gothic method.　　Stone of facade built up as wall in built.　　In Italy architecture is skin deep.　　In Italy it is a popular movement.　In France it is aristorcratic.

The French treated Renaissance as the Italians the Gothic. Merely the form, without understanding it.

In France Style after King.

(97)

Lecture 26, February 9, 1926.

Sources of Information   ①Nothing like Vasari.   No general source.

Students were much at loss to know definitely the history of the early period. ②No civic documents as in Italian cities.

ⓐ Had a set of document – accounts of the building of the king and also some specification.

ⓑ A book by du Cercau gives excellent information.

Palustre gives much credit to French men.   Before tendency was to give all to Italians.   He was first to give name of architects.   Italians he considered merely decorators. He was carried away by enthusiasm and invented details. for carrier of Stone masons.

Blunfield made careful study of documents and could not find prominence of stone masons over other artisans.   He arrived theory that king and nobles who ordered the building are accountable for the design.

In Du Cersau he states that the king was so well versed in building that it is impossible to call any body else the architect.   It is born in these early buildings some very rendom composition.

ITALIANS IN FRANCE

Francesco Laurana — Mentioned as being at court of Anjou 1460-1467. ~~Supposed to have built a part of Notre Dame.~~

Letters — Letters patent of Charles VIII mentioned **Fra Giocondo**, who is supposed to have built hte Pont Notre Dame. **De Cortonne** [Il Boccador] was mentioned in the same letter.

Paganio is credited with the Chateau de Panior, mentioned as working for court at 1530, money paid to him for making models. Had house at Blois. Was called to design Hotel de Ville at Paris, only building which can be proved to be by an Italian.

Justo worked in the early part of 16th century, credited with the Tomb of the Children of Charles VIII in Cathedral of Tour.

At Fontainbleau — French employed along with Italians. **Il Rosso** appeared in account for receiving money in supervicing building. **Primaticcio** appeared in account for stucco work for the Queen's chamber. **Serlio**, Blunfield thinks he designed wing of Bil Chemine. Also a house for the Cardinal at Farara. - only one doorway left.

SCHOOLS — In France architecture does not fall in to school. Nevertheless it is sometines classified as the following.

School of Loir — Light, picturesque grouping, graceful and delicate ornament. Examples: Blois, Chambord, Chateau at Tour.

School of Fon

(99)

<u>School of Fontainbleau</u>   Simplicity, lack of ornament.   Example: Chateau of Sr. Germain-en-Laye.   1540 Master Mason Pierre Chambiges.

<u>School of Burgandy</u>   Deeply undercut detail, ornament often out of scale, always original ad vigorous. Example: Archibishops Palace at Sens.

<u>School of South of France</u>   Big scale, sometimes much exeggerated.   Example: Maison de Pierre. Stone house at Toulou.   Details sometimes show Roman inspiration.   Bournazel is design quite Roman in character.

<u>School on the Coast</u>   Struck the mean between the South and Loirl Example: Hotel de Ville at Comte.

<u>Construction</u>   During the Middle ages, brick was little used.   During the Renaissance it was itroduced and used in combination with stone.   Revetment not used.   Stone and brick build in together as in Gothic practice.

In early Renaissance stone is entirely finished before put in place, this had been practice in the Middle Ages also. Rough cutting came latter.   Fluted pylasters.   Vaubts continued till the middle of the sixteenth century.   RRick frequently used for webbing.   Carpentry of hgh steep roofs. Disposal of water gargoil continued for sometime.

Lecture 27, February 15, 1926.

HOUSE OF VALOIS

Transitional Period

Charles VIII
Louis XII   Charles VIII was first of French Kings to go down campaigning in Italy.   He claimed Crown of Naples.   Was son of Louis XI, last of Gothic Kings.   Married to Anne of Britanny. 1483-1498

Succeeded by Louis XII, his cousin, who first married the widow of Charles VIII and later married Mary of England. Emblem is Porcupine.   It is frequently signed with Emblem and monogram.

Style   Charles VIII brought back a group of workmen whom he established at Ambois.   Architecture produced is French Flamboyant combined with Lombard Renaissance.   Buildings were really Gothic with sprinklings of Renaissance detail. The builders were French and decorators Italians.

Materials   Brick used combined with stone.   Brick used in various bond, thus interesting patterns.   Also majolica and flint.   Roof: tiles and shingles., Lead for lanterns

Scheme   Planning not changed much from Middle Ages.   Often use or build on Gothic Foundations.

(102)

<u>Louis' Wing
at Blois</u>   Or East Wing.   Brick and stone.   Stone as quoins at corners and windows.   Quoins irregular in shape and spacing.   Renaissance elements in ornaments.

<u>Chateau de
Gaillon</u>   By George of *Arbois*, Minister of Charles and Louis.   Incorporated parts of old palaces and used foundations of Gothic building.   Destroyed in revloution.   Fragments in court of l'Ecole des Beaus arts, Paris.   Known through du Cercau.

<u>Hotel de
Ville</u>   Orlean.   By Viart.   Great deal of brick used.

         <u>Interior</u> - <u>Ceiling of various</u> types.   Vault with pendentive.   Keystone extended below.   Carving elaborate.   Wooden barrel vault.   commonest form is to have the wooden beams exposed.   Often supported by heavy girders so as to break up span.

Hun ceiling panelled.   Floor, stone slap or tile.   Walls generally covered with tapestries, in some cases covered with wood panelling.   Window glazed only in impprtant rooms.   Somecases linen and oil paper used.

     Little evidence of eclessiastical building.   Church was reluctant to take up Renaissance.

STYLE OF FRANCIS THE FIRST

<u>Francis the First</u> Emblem is Salamandar, surrounded by flame, some out from mouth. Came to throne at auspicious moment. Man of great energy. In sympathy with Renaissance culture. Renewed Italian campaign and competed with Charles V of Spain for Imperial crown of Germany. Defeated and taken prisoner to Madrid.

Was tireless builder but lacked steadiness of purpose. *Plan*
<u>Scheme</u> Plans not much changd. On <u>Gothic buildings</u>. <u>Moat</u> and <u>drawbridge</u> continued. Greater tendency toward <u>symmetry</u>. Such at the Chateaux of Madrid and Chambord. <u>Stair still spiral</u> but not always towered. Straight run *Elevation* more common. <u>Pavilion appeared</u>. <u>Roof high</u>, sometimes half the height of facade. Sometimes flat paved roof, forming a terrace. <u>Chimneys and dorners elaborately</u> designed. Keeping general silhuoette of Gothic. Spindel for finials; scrolls for buttesses. <u>Windows square headed</u>, generally cut out by a nullion and transom near top. generally placed one above another. Framed by by pylaster. <u>No attempt to follow classical proportion of entablature.</u> Cornice served as sill of window above. They want in France the window as large as possible. Houldings frequently have Gothic profile.- cut out. Cap generally Corinthianesque. Francis I used

Wide Freize

his monogram almost on every thing.   Cantalabra shaft frequently used in place of order.   In many buildings spacings of widow irregular (Blois).

Materials   Ston and brck still used in same way.   Flint and slate as inlaid.   Also glazed terracotta, some of which came from Italy.

BLois
North Wing   North wing of Francis I.   Contains the elaborate staircase.   Irregular windw spacing. Lack of ~~space~~ relation in stair and facade.   Cornice is Gothic, used gargoil.   Stair case octagonal in plan, spiral, open to air. At angles are gothic buttresses cecorated with figures under canopies.   Continuous barral vault.   Out side facade later. Top story is open galler witheaves supported by cloumns. Author nknown.   Il Boccador had a house at this time in Blois.   Whether he had anything to do is no known.   Master mason is Philippi, but not necessarily accountable for the design

Lecture 28, February 16, 1926.

<u>Chateau de Chambord</u>   By Francis I, in Tourain, chentre of which is Tours. Near Blois. Accounts show that money was paid to Il Boccador for a certain model.

This is a hunting bos in the forest. Plan medaeval in character, but symmetrical - a block with four (4) round towers at corners. Inteior so arranged that at each corner there is a room, that is left is a great Greek cross, ceiling of elliptical vault. Center is double spiral staircases. Exterior is the same sort of thing as Blois. Windows one above another, pylasters framing it. Cornice as sill. Lower part is plain. Roof is picturewque, looms out from forest. Roof over large room is terrace. Towers covered by cones. Some of th domes two stories in height. Chimneys elaborate. Lantern over staircase is elaborate, it is the highest feature. Slate inlaid.

<u>Chateau de Fontainebleau</u>   Irregular and complicated plan. Cour Oval is where it began. Greater part by Frandis I, also long Gallery of Francis I, 192' long to connect building with entrance court (Cour de Cheval Blond).

<u>Wing of the
Beautiful Fireplace</u> — Blumfield thinks it is by Italians. Fontainebleau is severe. Stone unsuited to carving. No sculptural decoration. 2 stories with high roof. Simple chimneys and dormers. Divisions by pylasters, sometimes by quoins. Spaces between pylasters plastered or stuccoed. Interior had Italian influence. Il Rosso worked on stucco.

<u>Chateau of St.
Germain-en-Laye</u> — Near Paris. Originally a Gothic fortress. Has Gothic chapel by St. Louis (Louis IX). Gothic features largely destroyed during wars with England. Built around an irregular five-sided court. It is exceptional to have flat terrace roof, paved and supported on vault, thrust taken up by buttresses and tierods. Buttresses connected by arches at top. Treated with slender brick pylasters at corners and terminated by large urns. Wall spaces between are stucco or stone. Vertical strongly accented. Little ornament.

<u>Chateau de
Madrid</u> — Also by Francis I. Near Paris in Bois de Blon. Italian architect employed at least in design of terracota - Jerome della Robbia. Scheme is Two square blocks, with tower at corners connected

Lecture 29, February 23, 1926

<u>Hotel d'Ecoville</u>　　　At Caen.　　In facade of court. Instead of dormer over window below, it comes over pier.

<u>Interior of the Period</u>　　<u>Ceiling</u> still vaulted, ribbed - tradition of Gothic builders.　Such as at Blois Staircase and Chambord.　Stone slabs on arches - Azay-le-Rideau.　Commonest is open beam ceiling; room divided by large girders.　These ceilings decorative, beams often painted.　On surface often applied models in gilded plaster.　Panel cilings hung.

<u>Chimney Piece</u> large, hoods extend to ceiling, often richly decorated.　Emblem of builder often appear on hood. At Blois is salamanda.

<u>Door</u> in small panels, often treated with Gothic features, such as the linen-fold.　Also often filled with arabesques, an also architectural features.

<u>ECLISSIATICAL ARCHITECTURE</u>

With passing of feudalism, no more need for fortified dwelling.　But there is no such great change with church.

by hall.   Each unit with hip-roof.   Four stories in height. Two lowers with arcade.   Upper with pylasters.   Ornaments of terracotta.

<u>Villa at Moret</u>   Facade set up in Bank of Seine.   It is really a side of a court.   Used Cantalabra shaft.

<u>Chateau of Azay le Rideau</u>   A Private chateau.   By Betti.
Round towers at corners.   Staircase has straight run.   Ceilings form of flat arches. with Renaissance. detail

<u>Chateau of Chenonceaux</u>   On small island in the Chen [Seine].   Connected to Bank by Bridge.   General scheme is wide corridor with rooms on eight [either] sides.   Finer workmanship than any chateau by the King.   Later came into hand of Katherine Meddiddi.

①

(108)

Plan and section remained the same. Still involved Gothic principle. Even when in Classic, it uses only aa a matter of form - cloth a Gothic church in Renaissance detail. Renaissance is accepted only in form. The attitude of the French towards Renaissance is just like that of the Italians towards Gothic. This is not a period of church building.

<u>St. Eustache</u> One most important in church is St. Eustache. Paris. Authorship unknown. Facade and one bay of rl original destroyed. 5 aisle with chapels on sides. Largest in Paris next to Notre Dame - 243'x140' and 106' in height. Inner and outer aisles almost same geight. In principle it is a Gothic church in every way - in arrangement, principle, and plan. But orders used as decoration, regardless of proportion. Pylaster/s 15 diameters or more in height. Florentine arcade for trifol. Tracery is a kind of modified flameboyant. In a sense it violates good taste, but it has impressive interior. Naive, but not logical.

<u>St Michel at Dizon</u> Facade gelongs to this period. Used recess arches and dressed them up in Renaissance detilsa.

(109)

<u>St. Pierre
at Caen</u>   Chevet only.   Elaborately decorated.
                Arches elliptical.   Lady chapel has attic
with bull-eye windows, above which is ballustrade.

<u>Tomb of
George d'Amboise</u>   In Cathedral of Rouen.   Attributed
                to Rouland le Rou.   Consists of
a base which supported effigence of the deceased and that
of a nephew.   Above canopy are kneeling figures.

Lecture 30, March 1, 1926

STYLE OF HENRY II

Henry II
1547-59

Married to Catherine de Medeci, Italian Princess. Buildings marked either with ⊞ or ✕. He did not care much for her and came under influence of Dianne de Pointiers, his mistress.

This period also includes Francis II 1559-1560, Charles IX, 1560-74 and Henry III, 1574_89. This period may be calles the period of Cathrine de Medeci. Henry II was sone of Francis I. Was impressive, athletic, not very intellectual. Married at 14 to Catherine. Durig his reign, Italian claims given up. Tradition of kingdom improved.

In spite of th prosecution, Huguenots increased in number. Henry was accedentally killed in a tournament. Succeeded by son Francid II, who came to throne at 16. Married to Mary Stewart. His death gave regncy of Catherine over Charles IX. In reign of Henry III, civil wars, almost anarchy.

Catherine died few days after murder of brother of Charles. Henry assasinated 6 months later.

<u>Italian Influence</u>   School founded by Rrancis I had influence.

Italians brought had influence.   Architects sent to Italy.   Italian considered cultural. Italian models followed in literature.

Throne was Catholic, upholding that side of Pope and connected affiliation of Rome and Throne.

In spite of Italian influence, <u>had sense of construction</u>.

<u>Architect of the modern sense appeared</u>, who was responsible for design and building.

Characteristics

<u>Plan</u>   Regular, <u>symmetrical</u>, round towers replaced by <u>square pavilions</u>.   Stairs seldom placed in projecting tower.   Toward end of period, house became a rather fixed type; built <u>around rectangular court</u>, <u>units</u> are <u>main block</u>, <u>galleries</u> and <u>pavilions</u>.   Main block really one more than three sides, not like Italians that come on four sides. The fourth side by <u>gallery</u> one story high with terrace. Main block has a <u>basement</u>, <u>two stories</u> and <u>a roof</u> with a story.   <u>Pavilions</u> at angles and sometimes in center of facade, square in plan, project little over fadade. <u>Attic</u> often over and give prominence.

0

Orders   Use of orders increased.   Besides five orders, is a gahcec order.   Ionic with rustication.   Invented by de l'Ormel

With the period traditonal porportions became the practice. At end collosal order used, but more suitable in France, on account of big windows.   In some cases ignore order, often not corresponding with height of building.

Elevation   Greater vertical.   Roof high with curved profile.   Cornice more important than Italian.   Profile of mouldings became classic.   Less colour.   Majolica and brick not used.   Sculpture more massive.

Architects

① Lescot   Little known of his life. 1510-1578   Came from legal family.   Said to have neglected book to draw. Skillful in painting music and architecture.

② Goujon Jean GoujJean Goujon 1515-1568.   First heard at Rouen where part of organ gallery was attributed to him. Supposed also to have done someting on one of doors of S. Maclou.

In Cathedral of Rouen, Breze monument.

Fountain of Innocence   Not originally designed for type of thing as it is now.   Taken in 17 5 and rebuilt.

Rood Screen   or Jube S. Germain l'Auxerrois, Paris
Goujon and Lescot associated.

Louvre   The louvre is like a running commentary in the French Renaissance.

1527 Donjon of old Louvre was taken down.   1543 was competition for rebuilding.   Serlio the Italian approved the acceptance of Lescot's design and begun 1546 under Francis I.   Henry II confirmed appointment of Lescot and also by Frencis II and Henry III.

Scheme is square court.   175' on side.   About size of old Louvre.   Main block and at west and entrance on east. moat retained.   West wing and one bay of Southern wing only. First only a little corner of L.   Facade broken by slightly projecting pavilion, a new way β treating. Sculpture by Goujon.

Hotel Carnevalet   Also attributed to Goujon.   Now is Museum of History of Paris.

Chapel of St Valois   St. Denis.   Mousuleum of French King. Never Completed.   Destroyed 1719, Circular with Chapel.   Stone dome with wooden cupula above.

Lecture 31, March 8, 1926

DE L'ORME, PHILIBERT, 1515-1570

Born at Lyon, father contractor. Went to Rome, commissioned by Pope. Returned 1536 to Lyon. Arrival to Paris cate unknown. Appointed supervisor of buildings and fortifications in Brittany.

<u>Maison Henri II</u> at Rochelle is possibly by him. One of
<u>La Rochelle</u>
first acts of Henri II is to appoint him Architect to King and inspecter of Royal buildings. Unpopular. Did little important buildings. Completed Ball Room at Fontainebleau. Building is Francis I. Interior is this period.

<u>Porte Chapelle,</u>   is another of his works
<u>Compiegne.</u>

<u>Chateau d'Anet</u>   In 1532, employed to build this. Main building surroung three sides of court. Fourth side by low screen. On top of this is Nymph. Now in Louvre. On right side of court is Chapel. Circular dome. One of the earlier Renaissance domes in France. Either sides are subordinate courts. Incorporated old buildings. Now mainly destroyed. Main Pavilion is in court of Ecole des Beaux Arts.

1559 Henri II died. Catherine forced him to change Chateau Chennonceau. Dismissed. But took advantage to write.

Later employed on Chennonceau where he designed an app-orach, never carried out. Also banquet hall over bridge.

<u>Tuileries Paris.</u>   In 1564 A site of tile work near Paris was purchased. Intension is to develope into courts and gardens. Only partly done. Now known as Tuileris. It was to be a low building. Scheme in general is large court. Either side two elliptical courts. Only central portion carried out. In this he used French Order. Sdheme is One story. Arcaded high roof. and dormer. Pallissy designed garden. He was famous for ppottery.

He decided to carry along river a gallery to connect Louvre and Tuileries. Started by a wing, known as Le Galleries. Perpendicular to river, Grand gallery partly done.

Joan Bullant, 1515-1578.

Little known of arly life.   Stone mason.   Studied in Rome. Five manners of columns.

<u>Chatelot</u>
<u>Chantilly</u>   Collosal order.   Main entablature interrupted by windows.   Vigour in projection. Strong contrast of light and shade.

<u>At Ecouen</u>   Two portals in either sides of a court. Both used collosal order.   One of his pavillions forn a setting for Michael Angelo's Captere.

<u>Tuileries</u>   Upon death of de l'Orne, appointed architect to Tuileries.   Extended facade toward river.   Two stories and attic.   Used plain order.

<u>Hotle at Soisson</u>   Only part left.

<u>Anet Mauselleun</u>   attributed to him on account of similar design.

<u>Hotel Assezat,</u>

NICHOLAS BACHELIER 1487-1550

<u>Hotel Assezat
Toulouse</u>
Fine example of city house of the period.
Scheme: Open loggia with small story.
Brick and stone. Accross front is screen. Gallery on brackets. Two lower stories treated with arcade, inside of which are square head windows.

<u>Maison de Pierre
Toulouse</u>
Polygonal arches. Heavy ornament.

<u>Hotel Chanbellan
at Dijon</u>
Attributed to Hughes Sambin or Etienne Bruhe. Extremely heavy ornament

Lecture 32 March 9, 1936.

HOUSE OF BOUBON

STYLE OF HENRI IV

<u>Henri IV</u>   Bigins a new line.   In the period includes also Louis XIII.

Henry IV married Margaret of Valois, divorced and married Maria de Medeci.   His reigh marked a turning point in the History of the French.   Valois were individuals.   But with Henry IV came the idea of society.   Government established in Paris, court rarely left capital.

Classical studies increased.   In literature, Vitruvius studied in Italy.   Increased influence of protestants 'or Hugonots had tendency toward predestination - austerity.

Henry himself was a protestant but later bécame a Catholic.

The French Academyn was founded in 1634.   It made for uniformity.   Regulates the language.

Architecture looses picturesqueness.   But gains in ballande, order and unity.

Henry IV became king on assissination of his Cousin Henry III.   When he took hold of things, France was anarchy and it was his work to remove these.   With the Edit of

Monts, Protestants given permission. Henry takes king as the trust of the people.

He was assassinated in street in Paris.

<u>Louis XIII</u> His son, nine years old when came to throne. Maria de Medecci, his mother was regent. Cardinal Rechevoir was minister, brought to state a powerful position.

France during the period was torn by civil was, economy necessary. Brick used, because cheaper. Makes brilliant contrast with white stone.

This period in contact with <u>Holland</u>, country of brick, has influence. The austerity may be due to Protestant.

Architecture should be <u>a national expression</u> is idea of the French. <u>They seek to logic</u>. Expression of interior ~~on exterior~~.

Strong classical ideas came in <u>two channels</u>:

(1) <u>Through Catholics with Rome</u>;

(2) <u>Through Protestants</u> by idea of restoring primitive church.

(3) Also there is <u>Flemish influence.</u> In this period Rubens came, employed as court painter. A big Ruben room in Louvre now. Painted life of Maria de Medecci.

<u>Plans</u>   Changed little from previous period.   Fortified aspect lost more.   Moat retained without water, treated as sunken garden.   Pavilion in main block and independent gate pavilion retained.   Fourth side of court often enclosed with bulustrade instead of high walls.   House more open.

Rooms in series, pass through one to another.   But toward end of this a tendency of a little privacy, corridor introduced.   Less display on Street.   Lower story often with shops (city house).   Port cachier became only monumentale feature of facade.   Arch for coach to pass.

Orders less used.   Reserved for very monumental work and to accent inportant parts.   Relied much on brick wall with stone quoins in corners and chains in between. very frequently the quoins around windows connect with those below with chain.   Generally long and short.

Stalactite rustication used.

Roofs high, each unit in plan separatley roofed.   Some times profile curved.   Had also Mansard roof. Windows longer, often retained mullion and transon.   Vertical lines connected by chains, cedorated by quoins. Also bull eye windows.

Lecture 33, March 15, 1926.

Style of Henry IV

Authorship of buildings of this period cannot be determined with certainty.   However, there are two important architects families, both imployed on the Long Galleries, i. e.,

        Du Cerceau, and

        Metezeau.

The Du Cerceaus come from the Author of the Book and engraver.   The family tree as follows:

Jacque Androuet: 2 sons, 1 daughter;

Batiste   Jacque II   Julienne
(Charleval)   (a Pave de Flor)

Jean   Solomon de Brosse
(Hotel Sulley)   (Luxembourgh)

<u>At Fontainebleau</u>, three groups added:

(1) <u>Gallery des Cerfs</u>   Enclosure of the Queen's garden, conatins the Gallery, Brick and stone. (du Perac)

(2) <u>Baptistry</u>   Port de Fine, end of Cour Oval. Old gate puled down.   By Collin

(3) Cour des Offices. Both one story, dormers, higher adcents form pavilians. Walls plastered. Rustication and embalishments and cornices of brown stone. (Collin)

Place Royal (Chattillon) Henri IV interested in civic developments. Started to treat one place with one architectural scheme. He developed the Place Royal, now known as Place des Vosges. It was site of an old Hotel by Charles V. Catherine de Medecci had it pulled down. It is large park in center, and private hotels around. Red brick and white stone. Each is separate house, exterior is uniform. Quoins around windows and connected with chains, forming vertical accents. When first begun it was aristocratic, but fell down and the literary colony started. Victor Hugo lived there.

Hotel Vogue At Dijon. Private hotel, Nice arcade Loggia on inside of court. A't unknown.

St. Etienne du Mont By Biard. He was mainly a sculptor. Only Rood Screen in Paris churches.

Trinity Chapel at Fontainbleau has heavy details. By Freminet.

DE BROSSE Grand son of du Cerceau. It is not known whether he studied in Italy or not. Became the architect for queen. Most important work is the

Luxen-	Palace for Maria de Medicci. Was ordered
bourgh
	to copy Pitti Palace. He got inspiration
but two lookj nothing alike.   High roof changed character entirely.   Uses heavy rustication.   Scale is different.   Around three sides of court.   Fourth side is low gallery, with domed gateway in center. Main block three stories.   Superposed rusticated order. Garden also laid out by de Brosse.

LA MERCIER   Succeeded de Brosse as architect to King. Also Rechelier's architect.   1609-1613 in Italy. Worked on

Louvre.   Rechelier was Minister and decided to increase size of Louvre.   La Mercier enlarged the Court. Extended Louvre to four times of old Louvre.   In court is line marking old buildings.

Palais Royal   La Marcier also worked on Palais Royal, then
Royal
	Palais Cardinal. Much changed.

Sairway   rebuilt horseshoe stairway at Fontainbleau by Du Cerceau.

Facade st.	Architect unknown. Picturesque
Etienne de Mont
	Composition.

MANSART　　　Little known of his origin or training. Worked with de Brosse.　　Typified a classical spirit. Was vainmakes own geneolggy, as son kings.　　Trained horse with Rhymaic steps.

Blois
Orlean Wing　　　For Brother of Louis XIII.　　If carried out whole thing of Francis I will be destroyed.　　Design suggests Luxembourg.　　He bridges over to the period of Louis XIV.

Lecture 34, March 16, 1926.

Louis XIV   His Emblem is Sun; monogram .
The sun is sometimes called the sun of King.   Refered
to as Grand Monarch.   Period of reign 72 years.
France was then arbiter of Europe.   In literature
there were Molier, Racine, la Fontaine, etc.

Was son of Louis XIII.   Came to throne at five.
Anne of Austria regent.   Mazarin, minister.   Until
1661 when Louis XIV assumed bovernment himself.   1680
indication of decline.   Eidt of Nont caused Hugenots
to leave, is an aggressive foreign policy.   Failure
to understand need of people is germ of French Revolution.

Was tendency in part of state to direct influence
in artistic side.   Manufactury of Furniture established.
School of Design attached to factory.

1665, founded Royal Academy of Painters and Academy
of Architecture.   1666, French Academy in Rome.

Were publications of works on Theory of Architecture.
Francois Blondel belongs to this period.   He was adherent of Vitruvius.   Had desire to rid architecture
of its Baroque and extravagant tendencies.

Perrault is also important.   Thought general principles and proportion be taken from classic examples,
but

but taste and feeling of architect is final.　　Got away from archeology.

<u>Characteristics of the Period</u>　　Was <u>growth of Classic spirit</u>.
Was desire for display for splendor of domestic architecture.　　Was a rather a bombastic period.　　All these culminated ar Versaille, which became capital of France.

Toward <u>end of period, freer</u> tendency.　　Used freedom against formalism.　　Better conception of bigness and unity of classic composition.

<u>Decoration</u>　　Inspiration was Italy and Flanders, with increasing influence of Italy.　　Paladian with Baroque tendency.　　Free tendency but confine strictly in architecture.　　Temporary things set in!　　<u>Ceiling</u> beam concealed under hung ceiling.　　<u>Chimney</u> breasts concealed in wall.　　Much modelled stucco.　　Gilt metal fittings for caps and for balustrade of staircase.　　Coloured marble inlaid for chimney pieces, walls and pilasters.

Baroque tendency shows rounding off of corners. Use of shell on carouche at top panel to interrupt moulding. But in the main, architectural lines maintained.

Sun appears, face with rays. Gaily cork, military emblem.

Human figures used, robust in proportion, heavier. Lion, eagle, griffin used. Full leafy vetitation: oak, laurel, acanthus. Had an idea doing things like Roman. Painted composition as adjunt to architecture.

Le Pantre published books on architecture and ornamental compositions. Had great influence. Studied in Italy.

MONUMENTS

Louvre Le Vau completed court of Louvre. and Followed so far   . During the period this part was   carried out.

East
Facade Competition held for East facade. Two of the competitors were Mansart and Perrault, but none satisfactory. Was sent to Paussin who was then studying in Rome. He showed to Bernini, who condemned them. Bernini was invited to Paris, designed and accepted. Part of it carried until found could not be built. He left Paris in disgust. King then believed he preferred design of Perrault. Carried out and also a design to cover design which le Vau built.

This East facade is called Colonade of Loufre. 565' long. Consists a collonade of coupled, collasal Corinthian columns. Rest on high base, which is first story. Facade terminated in pavilians. Central pavilian has high pediment. These facde is merely decoration, do not express what is behind.

<u>Gallery of Apollon</u> Built over Pettite Gallerie by Lescot which serve as lower story. Upper story by Le Brun, who was director of Tapestry work. Was narrow long room, with Barrel vault, at end are cloister vault. No orders in interior, brokenup by paintings. In spite of elaborate decorations, architectural lines well accented.

<u>Tuileries</u> Facade entirely recased in this period by Le Vau.

Lecture 35, March 22, 1926.

<u>Versaille</u>   Originally hunting box of Henry IV, probably by de Brosse.   Brick and stone.   Under Louis XIV became residence of court.   Work begun by le Vau.   First was hunting box, building around three sides of court.   Very small structure.   Le Vau includes the enclosing of that on outside and extended two wings.   1646-1707, J.H.Mansart extended palace with two wings to north and south.   This Mansart is grandson of sister of Francois Mansart.   Also, in le Vau's scheme has lower facade, he built over with Large room, Gallerie de Glace. Decorated by le Brun, same man who did gallerie de Apolo. It has great barrel vault, architectural decorations are narrow long pylaster in green marble, with gild caps.   Walls white marble, walls filled in with mirrors.   Either end are square rooms, sall de l a Paix, and Salle de la Guerre. Also by le Brun.

Where terraces drop at the end is Orangerie, by Mansart, 1681.   Chapel, incorporated into one of the wings.

Garden facade monotonous.   Not bad in a unit, but whole thing is long.   From distance loose detail.   Central motf projects too far and cut off wings.

Fade on town side is confusing.   Looks almost like a little town.

134

Gardens famous, laid out by Le Notre, 1667.   Lawns, brooves, canals, summer houses, collonades, circles, fountains etc. Fountain is elaborate.   Path through dense wood.   Trees clipped At en is opening and fountain.   Effect is theatrical.

In ground of Park is le Grand Trianon, by J.H.Mansart.

Port St Denis.
By Blondel, who is more famous for Books.
Single arch, flanked by pier, with oblisks.

Palais Verdon
One of civic developments.   Open square, surrounded by uniform way.   Elongated octagon, with private house.   Treatment: Basement, collosal order pilasters with step roof with dormers.

Hopital des Invalids
To accomodate 6000 disabled soldies.
16 rectangular courts.   Little decoration. Is a chapel.   Most ornamental is effigee of gateway.   Dormer is breast plate and helmet.

Ecliasiastial

Two types:

(a) Basilican; churches, etc.

    1 – with or without trancept.

    2 – with or without dome.

(b) Radiating; chapels attached to institution.

    Dome dominating.

St. Roch and St. Sulpice both of Basilican type.   With trancept, ambulatory, chapel at spenig, lowedome, Barrel vault with penatration over nave.

Church of Surbonne

Church of Sorbonne — By le Mercier. One of the first Latin cross churc churches with done in France. Trancept wider than body of church. Dome at crossing is on axis of both front and side. Dome raised on high drum, and is covered on exterior by wooden domical roof which strings from a line a litte below crest of stone dome. Facade similar to that od De Jesuis, Rome.

Val de Grace — Francois Mansart, le Mercier and Le Muet. Usual plan, domical unit in denter. to which isn front is added nave, no Side aisle, but chapels. Nuns' gallery longer than other two arms. Central thing is square from outside with curious niches to fill up underneath pendentive. The wooden roof is a French feature.

Dome of the Invalides — 1693. JH Mansart. The name revers to the building itself. Built as a toyal chapel. Designed to dominate the hotel Placed directly back of chapel, so that if service is said in chapel, can be followed in dome. Finest example of radiating type. Square on exterior. Interior a Greek cross. Centre of which is octagonal. Piers splayed. Cornars filled wiht four octagonal chapels. Sanctuary which was added on to square is between ~~princixxixdnxx~~ the two chapels.

136

Principal come has drum supported on pendentives. The dome has a large occulus, covered by a second masonary dome. Still a thrid wooden. Windows on attic of the drum very clever. Now Napolion's tomb.

<u>Chapel of Versailles</u>   Incooperated into body of building, no facade.   Lower floors for servants, and upper arcade for the court.   Treated with arcade,   Main story is collosal supporting ~~on~~ a barrel vault.

图书在版编目(CIP)数据

梁思成图说西方建筑：汉英对照 / 梁思成著；林洙编. —— 北京：外语教学与研究出版社，2013.12（2019.4 重印）
ISBN 978-7-5135-3968-5

Ⅰ. ①梁… Ⅱ. ①梁… ②林… Ⅲ. ①建筑史－西方国家－汉、英 Ⅳ. ①TU-091

中国版本图书馆 CIP 数据核字 (2014) 第 005403 号

出 版 人　蔡剑峰
策划编辑　吴　浩　易　璐
责任编辑　张昊媛
特约编辑　孙　燕
装帧设计　视觉共振设计工作室
出版发行　外语教学与研究出版社
社　　址　北京市西三环北路 19 号（100089）
网　　址　http://www.fltrp.com
印　　刷　北京盛通印刷股份有限公司
开　　本　940×1080　1/12
印　　张　27
版　　次　2014 年 4 月第 1 版 2019 年 4 月第 5 次印刷
书　　号　ISBN 978-7-5135-3968-5
定　　价　198.00 元

购书咨询：(010) 88819926　电子邮箱：club@fltrp.com
外研书店：https://waiyants.tmall.com
凡印刷、装订质量问题，请联系我社印制部
联系电话：(010) 61207896　电子邮箱：zhijian@fltrp.com
凡侵权、盗版书籍线索，请联系我社法律事务部
举报电话：(010) 88817519　电子邮箱：banquan@fltrp.com
物料号：239680101

Palazzo Rucellai   In Florence Alberti designed the Rucellai Palace. It is important because it is the building where superposed orders were used in the forms of pilasters decorating *plintarium* the facade. The entablature is so designed that it serves as either an entablature or cornice. The projections here are all slight. Pilasters project 1/4 of the diameter. Doors are square. In this design the feeling of defense disappeared. Alberti designed

S. Maria Novella facade   S. Maria Novella facade for Rucellai. There were a number of tombs built in the walls. Church of S. Andrea was entirely

S. Andrea   free for Alberti to design. It is a Latin cross with a dome at the crossing and barrel vaults at the arms. There are no side aisles. Columns were placed at large and small spacing.

Rossellino   Rossellino (1409-1464) was associated with Alberti. They went to Rome on the invitation of St. Nicholas. They worked on St. Peter's. At Siena Rossellino built

Pallazzo Piccolomini   the Pallazzo Piccolomini. He was inspired by the Riccardi Palace. (Aeneas Silvius). Pope Pius the II. was the patron. He employed Rossellino to build a church, a palace and a town hall. The palace is isolated and built around a court.

S. Zacarria — The body of the S. Zacarria in Venice is more Gothic. The facade is entirely classical and has five stories. It is covered by a semicircular pediment. It does not express the building behind.

Scuola S. Rocco — Scuola S. Rocco has really two facades. The use of perspective as an adjunct to Architecture. Caisson ceiling was used in perspective and a colonnade in perspective.

Between Milan and Venice there are a number of cities where we have examples of architecture that show the mingling of the two forces. **Palazzo Consiglio in Verona** — In Verona, Palazzo del Consiglio was built by Giocondo who was a scholar went to France. When Raphael was working on St. Peter's Consiglio was also connected. The Palazzo consisted of a Florentine arcade terminated by pilasters. In Brescia, we find the early example of Roman arcade, and alogia for shops. The madona du la Miracoli used candelabra shafts. The scale is rather small. At **Palazzo Bevilacqua** — Bologna, Palazzo Bevilacqua is treated with fussy rustication on the exterior. The building is now in poor condition. The facade is like the Florentine. A big cornice crowns the building. The best part is the court of Florentine arcade. Enormous brackets are used to support a gallery as in Palazzo Fava. The second story is brought over the side walk. Porticos in Padua where you can

①